# Indian Names in Michigan

# Indian Names in Michigan

Virgil J. Vogel

Ann Arbor The University of Michigan Press

Published in the United States of America by
The University of Michigan Press
Manufactured in the United States of America

1997 1996 1995 1994 7 6 5 4

**Library of Congress Cataloging-in-Publication Data**

Vogel, Virgil J.
Indian names in Michigan.

Bibliography: p.
Includes index.
1. Indians of North America—Michigan—Names.
2. Names, Geographical—Michigan. I. Title.
E78.M6V64 1986 917.7'4'0014 85-13931
ISBN 0-472-10069-6 (alk. paper)
ISBN 0-472-06365-0 (pbk.)

*To Professor Sol Tax, who cares about Indians*

By the shores of Gitche Gumee,
By the shining Big-Sea-Water,
Stood the wigwam of Nokomis,
Daughter of the Moon, Nokomis.
Dark behind it rose the forest,
Rose the black and gloomy pine-trees,
Rose the firs with cones upon them;
Bright before it beat the water,
Beat the clear and sunny water,
Beat the shining Big-Sea-Water.

Henry Wadsworth Longfellow,
*The Song of Hiawatha*

# Preface

The term *place-name* as used in this book refers to the name of a geographic location that is sufficiently accepted to have a documentary existence in maps or written records. The name must have a degree of permanence. For our purposes, it must also be current, and so vanished names, except for a few unusually important ones, are here mentioned only casually, or as precursors of existing names.

Our names fall into two main groups, political and topographic. In the first group are the names of the state and its counties, townships, cities, towns, and villages, whether incorporated or not. We artificially define cities as places that have populations of more than 10,000, towns as places inhabited by 1,000 to 10,000 people, and villages as places having fewer than 1,000 residents. We also include the names of other public entities, including parks, preserves, and forests maintained by local, state, and federal governments. Because they do not fit our concept of place, or because they are impermanent or commercial, we exclude the names of bridges, buildings, camps, cemeteries, dams, farms, golf courses, highways, hospitals, industries, mines, monuments, neighborhoods, railway sidings, resorts, real estate subdivisions, and schools.

Topographic names are those given to all natural features of land or water. These include lakes, rivers, creeks, waterfalls, ponds, springs, marshes, bays, coves, channels, and straits, as well as capes, points, peninsulas, islands, rock formations, dunes, hills, mountains, prairies, and forests.

Our definition of *Indian name* is a broad one. For our purposes, it includes all names given to places by the Indians themselves, including translations of those names into European languages. It includes names of Indian origin given to places by non-Indians. Many names in this group were brought to Michigan by immigrants from other states and given to settlements in nostalgic remembrance of former homes. We also include those "Indian" names borrowed from legends or literature, even though some of them are pseudo-Indian and are so labeled. Also included are the bilingual and pseudo-Indian names manufactured by Henry R. Schoolcraft and others. By including them we can at least separate those of genuine Indian origin from those that are devoid of aboriginal roots. We also include names with Indian roots that are borrowed from countries of Latin America and the Caribbean. Some of these names entered historical literature or drama, and many entered English and other European languages, where they became so naturalized that their origin is forgotten. From literature came Lima and Capac, among others; while borrowed words used as place names include canoe, chocolate, hurricane, potato, and tobacco.

Last, we include some place names that are English or French but that commemorate historical or legendary events involving the aboriginal people. Such names as Battle Creek or Bloody Run record the aboriginal impact as much as the strictly Indian names do.

We have checked previous writers' interpretations of Michigan Indian names against both historical and linguistic sources. Frequently we have been compelled to disagree with earlier explanations and set forth our own views. Error, unfortunately, has a long life, for the accumulated copying of several generations of writers creates its own credibility merely from the weight of repetition.

In some cases no explanation, or only a tentative one, can be given, because of the degree of corruption of many names and the dearth of source materials for some. We, like our predecessors, are not immune from error, and so we invite those who have information either in support of or contrary to our views to submit it to the author in care of the publisher. We have tried to be comprehensive but not exhaustive. We invite readers to submit overlooked names to be considered in the event that this book is revised.

This work was aided by numerous people who answered letters, aided research, supplied materials, served as oral informants, or helped in other ways. The correspondents and oral informants are named in the bibliography, and we thank them for their courteous cooperation. Certain people have been of special help in locating sources and providing access to them. Among these are John Aubry of Chicago's Newberry Library; Mary K. Awdey of the Kent County Parks Commission; Sandra Clark, editor of *Michigan History*; John C. Curry, photo archivist for the Michigan Historical Commission; John R. Halsey, Michigan state archaeologist; Professor Alan H. Hartley of the University of Minnesota, Duluth; William A. LeBlanc, executive director of the Michigan Commission on Indian Affairs; Archie Motley, curator of manuscripts, Chicago Historical Society; Frank O. Paull, director of the Marquette County Historical Society; Professor Bernard Peters, Northern Michigan University, Marquette; Roger L. Payne, manager of the Geographic Names Information System, U.S. Geological Survey; Emelia Schaub, director of the Leelanau County Historical Society, Leland; Virginia Schaedig of Ocqueoc; and Edith M. Tuffts, president of the Kalkaska County Historical Society.

Particular thanks are due also to native American informants: Arnold Sowmick (Brown Beaver), chairman of the Isabella Chippewa Indian community, Mount Pleasant; Eli Thomas (Little Elk), of the Isabella Chippewa community; Father John Hascall, member of the Bay Mills Chippewa band and Catholic pastor at Bishop Baraga's former station, Assinins, Michigan; and Catherine Baldwin, Ruth Bussey, and Ardith (Dodie) Harris of the Grand Traverse band of Chippewa and Ottawa Indians, Peshawbestown.

I have also been aided by anonymous staff people of the Michigan Historical Commission and Michigan State Library, Lansing; the Harlan Hatcher Library of the University of Michigan, Ann Arbor; the Detroit Public Library; the Newaygo County Society of History and Geneaology, White Cloud; the Office of Anthropology, Smithsonian Institution; the Chicago Historical Society; the University of Chicago Libraries; the Northern Indiana Historical Society; the Royal Ontario Museum; Rand McNally and Company; and the public libraries of Northbrook and Highland Park, Illinois.

My wife, Louise, performed many mundane chores as research assistant and photocopier of endless sources in addition to this manuscript. Her maintenance of the household while I tried to organize a mountain of notes into a book is also an essential contribution. My son John chauffeured his father, assisted with photoduplication, and assumed the outside chores.

# Introduction

Place-names are linguistic artifacts. They are records of the past, conveying information about successive waves of inhabitants, native and European. They record people's perceptions of the environment. They tell us of the first people's spiritual and material culture, of historical happenings, and their description of the topography of land and water. Sometimes they identify fauna and land features that have since disappeared. As Emerson Greenman put it, "words are all that is left of earliest historic times besides the objects underground."[1]

The focus here is on native American place-names. In Michigan the great majority of these are the names of tribes and notable Indians and were adopted by whites. It must not be supposed that these names are simply a record of Indian occupation. A large number of place-names of Indian origin in Michigan, as elsewhere, were brought into the state by white immigrants from back East. These are a record of the places of origin of Michigan's settlers.

Indian names are a precious endowment, for they beautify and enrich the American map. They are fitting labels for our habitations and topography. They stand in contrast to ill-chosen Old World names that are records of a remote past only thinly connected with our national history. Indigenous names are a tribute to the innovative disposition that Americans have claimed for themselves.

Knowledge of the origin of our native names adds to the pleasure of travel and enlarges our understanding of the places where we live. They create a colorful and distinctive contrast to the tiresome parade of Springfields and Chestertons, of Athenses and Romes. As Walt Whitman put it:

> The red aborigines
> Leaving natural breaths, sounds of rain and winds,
> calls as of birds and animals in the woods
> syllabled to us for names. . . .[2]

There is no book-length account of Indian names in Michigan. There are only fragmentary articles, dealing mainly with the Upper Peninsula, and several county studies. Walter Romig's *Michigan Place Names* deals with the whole state, but lists only post office and township names, excluding topographic names, a class in which aboriginal names are most abundant. Its treatment of Indian names is inadequate.

The explanation of Indian names in most sources is incredibly chaotic. There are several explanations for many of the names. Too many writers have been satisfied with stories from folklore, and few have troubled themselves to examine early historical accounts and aboriginal language sources. Both are necessary in order to arrive at acceptable explanations of the origin and meaning of Indian names.

Even careful scholarship sometimes fails to produce a satisfactory explanation of some names. The reasons include the lack of sufficient early records, the degree of corruption of Indian names, the variety of spellings, the paucity in some instances of aboriginal vocabularies, the conflicting testimony of early accounts (as well as the interpretations of later investigators), the tendency of error to perpetuate itself, the magnitude of the task of dealing with names from several languages, the limitations of time, and the surfeit of writing on this subject by careless writers.

The present writer is a historian. He has spent much time studying maps and historical accounts in order to trace the evolution of Indian names. Relevant dictionaries and vocabularies have been examined. These, however, do not provide definitive answers to some name puzzles. It is first necessary to know in what language a name originated. Roots from different Indian languages may appear similar but have different meanings. Varying orthography is a problem; in Ojibwa there is no distinction between *b* and *p*. Some English phonetic values, such as the sounds of *l* and *r*, are absent. Surviving Indian names have gone through drastic changes to fit the phonetics of English or French. It is helpful to know something of the rules and grammatical structure of aboriginal languages.

Indian dictionaries list single words, but many Indian names are not single words; they are frequently phrases, often describing an action, condition, or characteristic. For such reasons, George R. Stewart observed, the testimony of early travelers who had contacts with the Indians is "more authoritative than any researches of modern linguistics can possibly be."[3] Unfortunately some early travelers, even educated ones, were not curious about Indian names, and certainly did not anticipate the future interest that might exist in such matters. Therefore, early accounts sometimes fail to illuminate a problem. It is a remarkable fact that not one early traveler thought to record the meaning of Wisconsin (Ouisconsin, Miskousing, etc.), and the question of its origin remains controversial, along with those of such Michigan names as Kalamazoo and Manistique. Despite the best efforts of scholars, the true origin and meaning of many Indian names is still unknown or uncertain.

This book is designed to interest average people, not experts. It avoids involved and pedantic disquisitions. It eschews the complexities of the international phonetic alphabet, for many Indian names were recorded according to no system, or phonetic systems devised by different individuals, or systems no longer current. Unravelling these differences is a task for linguists. In most cases we offer no pronunciation guide. Local pronunciation of a name, which we sometimes offer, is often in conflict with standard pronunciation, if such a thing exists, and is often at variance with the expected phonetic values that the orthography seems to indicate. Time and again, as I read off Indian names to Little Elk, he would say, "You are pronouncing it wrong." This, too, is a matter for linguists to deal with.

Our research has been supplemented by extensive travel throughout the state, which took us to sixty-three of Michigan's eighty-three counties and to five Indian reservations. By this means we were able to locate and interview local informants, including Indians, visit libraries and local historical societies, get a feel for the places we were writing about, and obtain a photographic record to enliven the book. Any place name researcher soon learns that information from local informants, like published data, is not always reliable, but it often leads to clues that point toward the solution. While we spoke with several Ottawa, Ojibwa, and Potawatomi Indians, we found that some of them have lost most or all knowledge of their language, and their information was limited. The main exceptions were Little Elk, an Ojibwa speaker at Isabella reservation, and Larry Matrious of the Hannahville Potawatomi community.

Although this work is as extensive as possible, given the limits of time and publishers' requirements for book length, it is not exhaustive. We have extracted all the current aboriginal names given in the computerized list of Michigan place-names issued by the United States Geological Survey, but have not been able to explain a few of them, as mentioned in chapter 21. Some of these may be aboriginal in appearance but are so corrupted as to be untraceable, or are in fact acronyms or some other variety of artificial name. Some, also, may be corruptions from European languages, especially French, which is normally spelled chaotically by Anglo speakers.

Because it was our intention to produce a readable book rather than a reference work, we have rejected the alphabetical listing technique of place-name dictionaries, and have instead chosen the topical method. We believe this is more meaningful and presents a better view of the range and significance of the names as historical and cultural records.

## Notes

1. Emerson Greenman, *The Indians of Michigan* (Lansing: Michigan Historical Commission, 1961), 10.
2. "Starting from Paumanok," stanza 16, in Walt Whitman, *Leaves of Grass* (New York: Penguin, 1943), 20–21.
3. George R. Stewart, *American Place Names* (New York: Oxford University Press, 1970), xii.

# Contents

I

# Michigan: The Name

It is common knowledge that the state of Michigan is named for Lake Michigan, which forms the western shore of the Lower Peninsula and the southern shore of the Upper Peninsula. The name is a modification of the words for "big lake" in Ojibwa and other Algonquian languages. The name is, in fact, a generic designation for any very large lake and was applied by one or more tribes, in one form or another, to all of the five Great Lakes.

One of its variant forms is the Gitche Gumee, "Big-Sea-Water," of Longfellow's *The Song of Hiawatha.* It is from the Otchipwe (Ojibwa) *kitchi-gami,* "great lake of the Ojibwa," or Lake Superior.[1] Applying the term to the lake now called Michigan, Louis Hennepin, the priest with LaSalle's expedition (1682) wrote that "It is called by the *Miami's, Mischigonong,* that is The Great Lake."[2] Father Ignatius Le Boullenger, in his manuscript "French-Illinois Dictionary" (1718) listed *Metchigamiȣi* as *grand lac* (great lake).[3] A nineteenth-century Ottawa, Andrew J. Blackbird, wrote that Michigan was pronounced *Mi-chi-gum* by his people, to whom the word meant simply "monstrous lake."[4]

According to Father William Gagnieur (1918), the term *kitchi* "comprehends every kind of greatness in quality," while *mitchi, michi,* or *missi* refers to quantity.[5] There is no substance to the claim that the name of Lake Michigan comes from that of the Michigamea tribe of the Illinois. They were named for Big Lake in Arkansas, near which they were first met by whites, including Marquette.[6]

Lake Michigan was the last of the Great Lakes to be discovered by Europeans. Champlain's map of 1632 does not show it at all. The first white man to see it was probably Jean Nicolet, who visited the Indians along Green Bay in 1634. No firsthand account of his trip has come down to us, and the narratives of others give no name to the lake. The early Jesuits, however, called Green Bay the Bay des Puans, or Puants (Stinkers), for the Winnebago Indians who lived there. (Their name, derived from their Algonquian neighbors, actually means, in this instance, salt-water people, because they came from the direction of Hudson Bay.) Some of the earliest mapmakers extended the name to the lake, which is called Lac des Puans in several maps made between 1650 and 1659, and one of the Jesuits, Paul Ragueneau, also gave the lake that name in 1648.[7]

Father Claude Allouez, the first French Jesuit missionary to the Ojibwa, founded missions at Sault Ste. Marie, Michigan, as well as at Chequamegon Bay and De Pere in Wisconsin. In 1666 he first became acquainted with Illinois Indians visiting his Lake Superior missions and apparently learned from them of the "lake of the Ilimouek, a large lake which had not before come to our knowledge." On its shores, he reported (1670), lived the Potawatomi, a group of whom were attacked by Senecas "at the foot of the lake of the Ilinoues, called Machihiganing."[8] That is the first reference to the name Michigan that this writer has discovered. Allouez's spelling is but one of dozens of variations that subsequently appear in maps and accounts. The *-ing* appended to it is simply a locative suffix, signifying "at" or "place."

On two so-called Jesuit maps from 1671 that are unsigned but probably the work of Father Allouez and/or Claude Dablon, as well as on Marquette's map (1673) and an undated map from this period by Jean Baptiste D'Anville, Lake Michigan is called "Lac des Ilinois." However, Marquette's companion in exploration, Louis, Jolliet, or Joliet,[9] showed it as "Lac des Ilinois ou Missihiganin."[10] This alternate name became Missiganin on a map by Hugues Randin (ca. 1681). According to the cartographic scholar Louis Karpinski, the first map to use the name Michigan in its present form was that by Melchisidek Thevenot, in 1681. There too it was coupled with "Lac des Ilinois" as an alternate name. Claude Bernou (1682) has "Mitchiganong ou Le Grand Lac des Illinois dit Dauphin" (Great Lake of the Illinois called Dauphin). This is, apparently, the first spelling of Illinois in its present form. Jean Baptiste Franquelin (1688) shows "Lac des Ilinois ou Michiganay." Louis de Louvigny (1697) followed with "Lac du Michigane ou Ilinois." Guillaume DeLisle (1703) attached the name "Lac Michigane" to Lake Huron, while the present Lake Michigan was simply labeled "Lac des Ilinois." At last, in 1718, DeLisle showed "L. Michigan" in its present spelling, with no alternate names, while to the east is shown "Lac Huron ou Michigane." These examples do not exhaust the various designations of these two lakes by the early mapmakers, but they are enough to show the evolution of the names.

Some of the older forms persist today in the names of lakes and streams in Michigan, Minnesota, and Canada. In Marquette County of the Upper Peninsula is Michigamme Lake, on which is situated the village of Michigamme. From the lake flows Michigamme River, which, flowing into Iron County, forms Michigamme Reservoir and another Michigamme Lake (this time with the term *lake* in second position), after which it joins the Brule River, forming the Menominee. Michigamme Township and Michigamme State Forest in Marquette County are named from this complex of waters.

Other variants are Michigamaw, named by the Plains Cree, in northern Saskatchewan; Michikamau Lake, from the Naskapi, in Labrador; Kamichigamaw Lake, a source of the Ottawa River in Quebec; and Kitchigama River, a tributary of Rupert Bay of James Bay in Quebec.[11] Also related to these names are those of Kitchi Creek and Lake in Beltrami County, Minnesota, and Kitchi Creek in Itasca County of the same state.

Michigan became the name of a political subdivision for the first time when the Michigan Territory was created in 1805. On January 26, 1837, Michigan, reduced in size from territorial limits that included Wisconsin, became the twenty-sixth state.

In Michigan several local names are derived from the state name. Michigan Center, in Jackson County, was so named because it was presumed to be located midway across the state, between the eastern and western extremities of the Lower Peninsula. Michiana Shores, Berrien County, is named for its proximity to the Indiana boundary, and Michillinda, in Muskegon County, was named by resort owners for Michigan, Illinois, and Indiana.[12]

The name of Michigan has spread through ten states, from New York to Alaska, and to Canada. In some cases, such as Michigan City, Indiana, the places are named for the lake. Most of the others were apparently named by expatriates from the state. These include Michigantown in Clinton County, Indiana; Michigan Island, Wisconsin, in the Apostle Islands of Lake Superior; Michigan Creek, in Steuben County, New York; Michigan City, in Benton County, Mississippi; Michigan Valley in Osage County, Kansas;

Michigan Centre in Alberta, Canada; and Michigan in Nelson County, North Dakota. The last was once called Michigan City, for the native state of the early settlers. Because it was confused with Michigan City, Indiana, a whole trainload of iron ore intended for the city in Indiana was sent to North Dakota instead. The post office and the inhabitants prudently deleted "city" from the name.[13]

But this is not all. Prospectors for gold planted the name of Michigan in three more states. They named Michigan River, a tributary of the North Platte, as well as North Michigan Creek and Michigan Lake, all in Jackson County, Colorado. Similar adventurers introduced the names of Michigan Bluff, in Placer County, California; Michigan Bar in Sacramento County; and perhaps Michigan Flat in Lassen County of the Golden State. Finally, about 1898, Michigander gold seekers gave the name of Michigan to four creeks in Alaska.

## Notes

1. Frederic Baraga, *A Dictionary of the Otchipwe Language, Explained in English* (reprint, Minneapolis: Ross & Haines, 1966), 1:153.
2. Louis Hennepin, *A New Discovery of a Vast Country in America* (Chicago: A. C. McClurg Co., 1903), 1:62.
3. Original in John Carter Brown Library, Providence; copy in Newberry Library, Chicago; citations on pp. 108, 160.
4. Andrew J. Blackbird, *History of the Ottawa and Chippewa Indians of Michigan.* (1887; reprint, Petoskey: Little Traverse Regional Historical Society, 1977), 93.
5. William F. Gagnieur, "Indian Place Names in the Upper Peninsula, and Their Interpretation," *Michigan History* 2 (July, 1918): 548.
6. Frederick W. Hodge, ed., *Handbook of American Indians North of Mexico,* Bureau of American Ethnology Bulletin no. 30 (Washington, D.C.: U.S. Government Printing Office, 1907–10), 1:856.
7. Louis C. Karpinski, *Historical Atlas of the Great Lakes and Michigan* (Lansing: Michigan Historical Commission, 1931), 8, 9, 12; Reuben G. Thwaites, ed., *Jesuit Relations and Allied Documents* (Cleveland: Burrows Brothers Co., 1896–1901), 33:148–51.
8. Louise P. Kellogg, ed., *Early Narratives of the Northwest, 1634–1699* (reprint, New York: Barnes & Noble, 1967), 123, 153.
9. Eminent authorities are divided on the proper spelling of the name Joliet. Since one *l* is used in his name on the map, we follow suit.
10. All map references here and below are from Karpinski, *Historical Atlas,* and

Sarah Jones Tucker, *Indian Villages of the Illinois Country*, vol. 2, *Scientific Papers*, pt. 1, *Atlas* (Springfield: Illinois State Museum, 1942).

11. Bernard Assiniwi, *Lexique des Noms Indiens en Amérique* (Ottawa: Editions Lemeac, 1973), 51, 74–75.
12. Walter Romig, *Michigan Place Names* (Grosse Pointe: Walter Romig, n.d.), 366.
13. Federal Writers Program (FWP), *North Dakota: A Guide to the Northern Prairie State* (Fargo: Knight Printing Co., 1938), 240.

# II
# Tribal Names

Most of the Indian names borne by Michigan's cities, counties, lakes, and rivers are those of Indian tribes or individuals. Generally these places were named by whites who paradoxically honored the natives while despoiling them. Settlers named places not only for the resident tribes, but also for tribes in the states they had left and even for tribes in the West that they had never seen.

The tribes native to Michigan, with one exception, belonged to the Algonquian language family. These included the Menominee and the loose confederation called the Three Fires—Chippewa (or Ojibwa), Ottawa, and Potawatomi. In earlier times, Michigan was the home of Miami and Sauk Indians. The remaining major tribe was the Huron, later called Wyandot, which belonged to the Iroquois language family even though they were mortal enemies of the six Iroquois nations of New York.

## The Algonquin

The Algonquian family is named for the Algonquin, a small tribe that encompassed several bands named for their localities, whom Champlain found living along the St. Lawrence and Ottawa rivers in the early 1600s. They still live on several reservations and in communities in Ontario and Quebec.[1]

In Michigan the name Algonquin is that of a lake and village in Barry County, parks in Gogebic and Wayne counties, and a waterfall in Black River, Gogebic County. In the 1830s Henry R. Schoolcraft chopped syllables from the name and joined them to others of his choice to manufacture such names as Algansee, Algoma, and Algonac, which will be examined later.

There is disagreement about the meaning of Algonquin, partly caused by the variety of forms in which the name has appeared, starting with *Algommequins* on Champlain's map. Schoolcraft decided that it meant "people of the opposite shore." William W.

Warren, Ojibwa historian of the last century, gave the name of the "Algonquins proper" as *O-dish-quag-um-eeg,* "Last water people." But that does not appear to be a definition of Algonquin. J. N. B. Hewitt, Seneca ethnologist, thought that Algonquin was derived from a Micmac term, *algoomaking,* "at the place of spearing fish."[2]

## The Chippewa (Ojibwa)

The Chippewa were from the early days the most numerous tribe in the land that became Michigan, and remain so today. In the Upper Peninsula they occupy reservations at Keweenaw Bay, L'Anse, and Bay Mills, in addition to communities at Lac Vieux Desert, Sault Ste. Marie, and Sugar Island. In the Lower Peninsula the Saginaw band lives at Isabella reservation near Mount Pleasant in Isabella County, and at Saganing in Arenac County. Other Chippewa reservations are found in Ontario, Wisconsin, Minnesota, North Dakota, and Montana. The language of this people was the *lingua franca* of the upper lakes, and they were early allies of the French.

Chippewa is a corruption from Odjibwa, Ojibwa, Ojibway, or Otchipwe, to list but a few of the dozens of variant spellings of this name. Schoolcraft, often fanciful and contradictory, remarked in one place that Chippewa was anglicized from an original term referring to "the power of virility." Elsewhere he wrote that Odjibwa signified "sibilant or hissing voices."[3] Neither view has won any acceptance. More popular is the view of William Keating, narrator of Long's expedition to Minnesota, who wrote in 1823 that "it signifies plaited shoes, from the fashion among these Indians of puckering their moccasins."[4] The most generally accepted meaning is "to roast till puckered up," referring, as Keating indicated, to moccasins. A more startling explanation was given in the last century by Ojibwa historian William Warren, who said that an element of the original name, Ab-boin-ug—"roasters"—arose from the custom of burning Dakota captives.[5]

Chippewa became a popular place-name in Michigan. Chippewa County, bounded on the north by Lake Superior, on the east by the St. Marys River, and on the south by Lake Huron, is the most visible toponymic memorial to the first inhabitants of that region, whose descendants still live there. In the county is Chippewa Township. Elsewhere in the Upper Peninsula is Chippewa Point on Big Bay de Noc, Delta County; another Chippewa Point in Chippewa County; Chippewa Hill and Falls in Gogebic County; and Chippewa Harbor on Isle Royale. In the Lower Peninsula are Chippewa townships in Isabella and Mecosta counties, Chippewa Lake and the village named for it in Mecosta County, and the Chippewa River, which flows from the lake, crosses Isabella County, and joins the Tittabawassee River at Midland in Midland County. Chippewa River State Forest is named for it. There are also Chippewa creeks in Osceola and Mecosta counties, Chippewa Bank in Chippewa County, and the communities of Chippewa Beach in Cheboygan County and Chippewa Vista in Mecosta County.

Ojibwa is the alternate and more nearly correct name of this widespread tribe. Lake Ojibway (thus spelled) and Mt. Ojibway (1,136 ft.) on Isle Royale, Keweenaw County, and Ojibway Island in the Saginaw River at Saginaw are the only examples of it on the map of Michigan.

## The Ottawa

Closely allied to the Ojibwa were the Ottawa. In early times they were often linked with the Algonquin of the Ottawa River, in Canada, which has their name. They were driven into Michigan by Iroquois raids in the mid-1600s and settled in both peninsulas. When white settlement began they were mainly concentrated around Grand Traverse and Little Traverse bays and northward to the Straits of Mackinac. Their largest settlement was L'Arbre Croche, between Good Hart and Cross Village, Emmet County. Efforts to remove them in the nineteenth century caused many Ottawa to flee to Manitoulin Island of Georgian Bay, part of their old Canadian homeland, where they remain. Other Ottawa were removed to Kansas (where their name is on a city and county) and then to northeastern Oklahoma (where their name is also on a county), while a sizable number, then without a reservation, continued to live in small settlements near Harbor Springs, Peshawbestown, and Good Hart.

In 1878 Father Baraga interpreted Ottoway as "an abbreviation of *ottawokay,* his ear, or *otawask,* and *watawask,* bull-rushes [*sic*], because along the river there are a great many of those bullrushes."[6] In 1901 Father George Lemoine thought Ottawa was derived from Montagnais-Algonquin *uteuau, uteu,* "it is boiling," referring to the Chaudiere Falls in the Ottawa River.[7]

The views of Baraga and Lemoine are generally rejected, while that of William Warren has won acceptance. He called the Ottawa O-dah-waug, "trading people," and explained:

> The Ottoways remaining about the spot of their final separation, and being thereby the most easterly section [of the three associated tribes], were first discovered by the white race, who bartered with them their merchandise for furs. They for many years acted as a medium between the white traders and their more remote western brethren, providing them in turn at advanced prices, with their most desired commodities. They thus obtained the name of Ot-tah-way, "trader," which they have retained as their tribal name to the present day.[8]

The Ottawa are honored in Michigan by the names of Ottawa County (where the county seat is Grand Haven); Ottawa Creek and the villages of Ottawa Beach and Ottawa Center in Ottawa County; the Ottawa River, Ottawa Lake, and a village of that name in Monroe County; Ottawa Lake in Iron County; and Ottawa National Forest, embracing several counties of the Upper Peninsula. The capital city of Canada is named for them, along with places in five American states. According to some, the name of Tawas on several places in Iosco County is an abbreviation of the tribal name Ottawa, but the evidence suggests it is from a personal name.

## The Potawatomi

The third member of the Three Fires were the Potawatomi, whom the French first found living in the region around Green Bay and southward on or near the shore of Lake Michigan. They later spread about the southern end of Lake Michigan, occupying northeastern Illinois, northern Indiana, and much of southern, and especially southwestern, Michigan, below Grand Haven. Most of the Potawatomi were removed in the 1830s, some to western Iowa and northwest Missouri, others to eastern Kansas and finally to central Oklahoma. Stragglers and refugees who escaped the net survive today at the Hannahville Reservation in Menominee County and in small settlements near Athens in Calhoun County (on a state reservation), Dowagiac in Cass County, and Hartford in Van Buren County. There is also a Potawatomi reservation in Forest County, Wisconsin, and others are in Canada at Walpole Island near Detroit.[9]

On the meaning of Potawatomi there is no serious debate; it is generally linked to fire. William Warren gave the name as *Po-dau-waud-um-eeg,* "those who keep the fire." He explained that at the time of the separation of the three associated "fires":

> The Potta-wat-um-ees moved up Lake Michigan, and by taking with them, or for a time perpetuating the national fire, which according to tradition was sacredly kept alive in their more primitive days, they have obtained the name of "those who make or keep the fire," which is the literal meaning of their tribal cognomen.[10]

William Keating, who met Potawatomi near Fort Wayne, Indiana, in 1823, interpreted the name nearly the same—"we are making fire"—but told a different story of its origin. He was told that a Miami Indian met three Indians of a strange tongue and invited them to his cabin. During the night, two of the strangers left, while one of them, with the host, remained asleep. The departing pair took embers from the cabin and made a fire near the door; when the remaining pair saw it they understood it to signify a council fire in token of peace between the two nations.[11]

The Potawatomi are commemorated by several personal names on the Michigan map, but the tribal name survives on but four insignificant places: Potawatomi Falls in the Black River, Gogebic County; Potawatomi Island near Seul Choix Point in Schoolcraft County; Pottawatamie Bayou in Ottawa County; and Potawotamie Lake, a name given to the outlet of Galien River, Berrien County.[12] Three spellings occur, but Potawatomi is the form approved by the former Bureau of American Ethnology.

## The Noquet

On September 28, 1698, the Jesuit missionary Jean Francois Buisson de St. Cosme wrote that upon approaching "The Bay of the Puants" (Green Bay), "one passes on the right hand another small bay called that of the Noquest. The Bay of the Puants is inhabited by several savage tribes: the Noquest, the Folles Avoine [Wild Oat, Menominee], the Renards [Foxes], the Poûtoûatamis, and the Saki." In 1775 Jonathan Carver wrote that on the northwest part of Lake Michigan the waters branched into two bays, called Bay of Noquets and Green Bay.[13]

The Noquet had been reported earlier on the shores of Lake Superior. They have not been positively identified, and it is believed that they were adopted by the Ojibwa or the Menominee. Their name is held to mean "bear foot," another name for the bear gens of the Ojibwa.[14] Today Noquet has been shortened to Noc and appears in the names of Big Bay de Noc and Little Bay de Noc, as well as Bay de Noc Township, Delta County.

## The Menominee

On the sixth of May, 1670, Father Claude Allouez reported that "I paid a visit to Oumalouminek, eight leagues distant from our cabin, and found them at their river. . . ." The river indicated is today called Menominee River, for the tribe.

Three years later Father Jacques Marquette, entering Green Bay, wrote that "The first nation that we came to was that of the Folle Avoine. I entered their river. . . ." He explained that "The wild oat, whose name they bear because it is found in their country, is a sort of grass, which grows naturally in the small rivers with muddy bottoms, and in swampy places." He gave a detailed account of how the grain was harvested and prepared as food, and remarked that "Cooked in this fashion, the wild oats have almost as delicate a taste as rice when a better seasoning is added."[15]

Today's name for the "wild oats" is wild rice (*Zizania aquatica*), and it has become a gourmet food.[16] The Menominee tribe today occupies a reservation in Menominee County, Wisconsin. Their name signifies "wild rice people" in the several forms found in Algonquian languages. Warren called them "O-mun-o-min-eeg, wild rice people," in Ojibwa. In Potawatomi wild rice is *man-o-min*; in Menominee, *mänóme* (rice) plus *inän'ni* (man) forms the tribal name.[17]

The Menominee left their name on their river, which forms part of the boundary between Wisconsin and the Upper Peninsula of Michigan; on Menominee County, which borders the river; on the city of Menominee, county seat, and its township, on the Menominee Iron Range; and Menominee State Forest.

## The Miami

At Fort Wayne, Indiana, where the Miami Indian village of Kekionga once stood, the St. Joseph River (one of three with that name in the same region) unites with the St. Mary's River to form the Maumee, which joins Maumee Bay of Lake Erie at Toledo, Ohio. North Maumee Bay, an arm of Maumee Bay, is within the limits of Monroe County, Michigan. Maumee is a variant form of Miami, and was given to this river, formerly called Miami of the Lake, to distinguish it from another Miami River that joins the Ohio River near Cincinnati. Their only other topographical presence on the Michigan map is in the unincorporated communities of Miami Beach, Cheboygan County, and Miami Park, Allegan County.

The Miami occupation in Michigan was transitory. First seen by Europeans in the neighborhood of Green Bay, Wisconsin, the Miami migrated southward along the west shore of Lake Michigan after 1650 and had a village at the site of Chicago for a few years following 1695. At the same time, they were established along the Kalamazoo River and St. Joseph River (not to be confused with the river of the same name in Indiana). LaSalle and others called the St. Joseph River of Lake Michigan the river of the Miami. In 1703 the Miami had a settlement at Detroit, but during the eighteenth century they settled along the Wabash River and its tributaries in Indiana. From there most of them were removed in the 1840s, first to Kansas and then to Oklahoma.

The name Miami has been variously interpreted as "people of the peninsula" (*omaumeg*) and "pigeon." According to C. C. Trowbridge (ca. 1830), the Miami term for pigeon was *meemeea,* which grew by use into *meearmee.* The name Miami in Florida is not related to the name in Michigan, Indiana, and Ohio.[18]

## The Sauk

Several writers maintain that Saginaw Bay is named for the former presence of the Sauk Indians, an Algonquian tribe closely allied to the Foxes. If they lived there, it was before the arrival of the whites.

Saginaw will be discussed elsewhere, for there is much evidence for concluding that it is from Ojibwa *sâging,* "at the mouth." The proper rendition of the name of the Sauk is Osauki-wug, "people of the yellow earth."[19] In place names, however, it is usually shortened to Sac or Sauk, although the older form is approximated by the names of Ozaukee County in Wisconsin and Osakis Lake in Minnesota. In Michigan, Sac Bay on the south end of the Garden Peninsula, Delta County, has the tribal name in the form ordinarily used by the French. The alternate spelling in seen in Sauk Head, a hill, and Sauk Head Lake, Marquette County, and the Sauk River, Branch County.

## The Huron

The only non-Algonquian tribe in Michigan during recorded history was the Huron, whose language was Iroquoian. There are more places in Michigan named for the Huron than for any other tribe. Before 1649 these Indians lived on the north shore of Georgian Bay, but in that year the Iroquois began the attacks that led to their dispersal, and most of those who escaped massacre fled into Michigan, although some found their way to Quebec. The majority first went into the Upper Peninsula, and eventually some settled in the Detroit area and around Sandusky Bay, Ohio. In that vicinity they became known under their own name, Wyandot, derived from *wendat.*

The earliest geographical use of the word Huron is in the application of Huronia to the "Huron country" in the *Jesuit Relations* of 1634–36.[20] Father Paul Le Jeune, explaining the name in the *Jesuit Relations* of 1639, wrote that some of these Indians wore their hair in ridges, and because of the appearance of their heads to some soldier or sailor, they were called *hures* (boars), from which came the name Huron. That explanation was repeated in later years by Fathers Francisco Bressani (1653) and Pierre F. X. Charlevoix (1721) and has been accepted by most American writers ever since. J. N. B. Hewitt pointed out that the word *huron* was in use in France as early as 1358 "as a name . . . signifying approximately an unkempt person, knave, ruffian, lout, wretch."[21] It still appears in French dictionaries.

Despite all this, the case for French origin of this name as applied to the tribe is based on guesses, and several questions remain unanswered. The first Frenchmen to see the Huron, Jacques Cartier (1534) and Samuel de Champlain (1603), never used the word Huron. To Cartier they were simply *sauvages,* "wild men," and to Champlain they were *Bons Iroquois,* "good Iroquois." Why would the Huron be uniquely named for the scalp lock, which was common in many tribes? Why, moreover, would they be saddled with a contemptuous name when they were friendly allies of the French? It is possible, even likely, that Le Jeune's explanation is simply a folk etymological explanation of a gallicized native term.

In 1885 Russell Errett held that the name Huron came from Irri-ronon, the Iroquois name of the Eries, or "Cat Nation." The French commonly placed an *h* before vowels, often writing *Hiroquois* for Iroquois and *Hohio* for Ohio. So also, it seems, they changed *Irri-ronon* to *Hirri-ronon* and then abbreviated it to *Hirr-on* and *Huron.* In Errett's view, Huron was essentially the same name as Erie, differing in spelling only because the name Erie came directly from the Iroquois, while Huron came directly from the French. He believed that both names represented a tribal totem.

There is support for his hypothesis in the fact that the Neutral Nation of Canada, commonly regarded as a Huron subtribe, was called "Cat Nation" by the Iroquois, according to Morgan. Moreover, Charlevoix called the Erie "an Indian nation of the Huron language." Finally, after being driven from Georgian Bay, some Huron joined the Erie.

Another challenge to claims of French origin of Huron came from William M. Beauchamp (1907), who called that version a "fanciful story." He pointed out that "the Hurons were not known to the French by this name for some time. It seems to have been used only after visits to their country, and is probably of aboriginal origin." He concluded that Huron was "strongly suggestive of the Huron-Iroquois word ronon, a *nation.*"[22]

The Huron are best known for having their name attached to one of the Great Lakes. At first, Champlain called the second largest of these lakes *Mer Douce* (freshwater sea), and others called it *Michigane* (see chap. 1). Father Hennepin called it "Lake Orleans or of the Hurons." By the middle of the eighteenth century the lake was called Lake of the Hurons, at least as an alternate name.[23] It was also called *Ottawa kitchigami* by the Algonquin, and *Caniatare* by the Iroquois. Lac des Hurons or Lac Huron was commonly used by the late 1600s and soon eliminated all other names.

Some of the places named Huron are named from the lake, but in Michigan most of them are named for the one-time presence of the tribe. The exceptions are a few places in the thumb of Michigan, in Huron and Sanilac counties. In the Upper Peninsula, to which the Huron first fled when driven from Canada, their name has been given to a small lake in Houghton County; the Huron Mountains in Baraga and Marquette counties; the village of Huron Mountain in Marquette County; the Huron Islands in Lake Superior, Marquette County; Huron Bay of Lake Superior in Baraga County; the Huron River, a tributary of Lake Superior in Baraga County; Huron River Point at the river's mouth, just over the line of Marquette County; Little Huron River in Marquette County; and the village of Hurontown in Houghton County. Another Huron Bay is on the south shore of Drummond Island, in Lake Huron, Chippewa County.

In the Lower Peninsula are Huron Beach, a village in Presque Isle County; Huron Bay on the lake shore in Alpena County; Huron County (where the county seat is Bad Axe); Huron City in the same county; Huron townships in Huron and Wayne counties; Huron National Forest in Oscoda County; the city and township of Port Huron and village of Huron Beach in St. Clair County; Huron Gardens, a suburb of Pontiac, and Huron Swamp, in Oakland County; Huronia Heights in Sanilac County; and the Huron River, a tributary of Lake Erie, in Washtenaw, Wayne, and Monroe counties. The name Huron is found also in nine other states.

## The Wyandot

In the *Jesuit Relations* of 1639, Father Paul Le Jeune described the Huron as an alliance of four tribes, adding that "the general name, and that which is common to these four Nations in the language of the country is Wendat."[24] From *wendat* came the better-known name Wyandot, or Wyandotte. These are the names given to the Huron remnants that settled near Detroit and Sandusky Bay in Ohio after 1650. Louise Kellogg believed that the name Wyandot was properly limited to the "Tionnotates, the Petuns or Tobacco Nation," and that the Hurons of Lorette, near Quebec city, are "the only descendants from the true Hurons of Champlain's time." However, Father Le Jeune's comment suggests a broader application. Moreover, as the name Huron vanished from use on the

American side of the border, while remnants of the various Huron subtribes survived there, it seems likely that they were all embraced under this name. Elisabeth Tooker, scholar of the Huron, wrote: "The Huron, or *Wendat* as they called themselves, were a league of four nations [tribes] sharing a common language, but each retaining its own traditions." The meaning of Wyandot, according to Swanton, is "perhaps 'islanders,' or 'dwellers on the peninsula.'" Jacob P. Dunn held that the name means "people of one speech."[25]

Wyandot living on the Huron River near Detroit and at Sandusky Bay were removed in 1842 to the site of Kansas City, Kansas, and in 1867 to northeastern Oklahoma. In Michigan their name survives in that of the city of Wyandotte, in Wayne County, south of Detroit, where a Wyandot village stood until 1818. Their name appears again in Wyandotte County, Kansas, and the town of Wyandotte, Oklahoma, a cave and village called Wyandotte in Crawford County, Indiana, and a creek and village in Butte County, California. The orthodox spelling, Wyandot, used by ethnologists, is applied to a county in Ohio.

## The Puzzle of Nottawassepee

In Calhoun County, the Nottawassepee River flows from a lake of that name into St. Joseph County, where it meets the Rock and Portage rivers to form the St. Joseph at Three Rivers. (This St. Joseph River flows into Indiana to join the Elkhart, and should not be confused with two others of the same name.) From Nottawassepee River were named Nottawa Prairie, the township of Nottawa, and the villages of Nottawa and Wasepi, all in St. Joseph County, as well as Nottawa Creek, Drain, and Lake in Calhoun County.

The word *river* attached to this name is redundant, for *sepee* means river. There are several stories about this name. One holds that it means "river of the Ottawa." That is unlikely for two reasons. First, the Ottawa did not live there, and second, the presence of *n* cannot be explained thereby. It is common to drop initial sounds from Indian names, but the reverse seldom happens. Another view holds that the name comes from "a St. Joseph band of Indians known as Nottowa-Seepees." These people, who were Potawatomi, were surely named for the river, not the river for them. Still another writer claims that the river was named for a Potawatomi chief. A Potawatomi chief called No-ta-way was reported living in Calhoun County in the 1840s, and another individual of the same name lived on the Ojibwa reservation in Isabella County, where a township bears his name. However, unacculturated Indians did not name features for individuals; furthermore the place-name was in earlier use.

As mentioned elsewhere, *nottawa* means "enemy" or "adder" and was sometimes applied to a species of rattlesnake. It was a common Algonquian epithet for the Sioux, Iroquois, and even the friendly Wyandot, because of their Iroquoian speech. Territorial governor Lewis Cass (1826) called "Nautowa Sepe" the "River of the Wyandots." This appears to be the most plausible explanation of the origin of the name of this river.[26]

## The Erie

The Huron-Wyandot were not extinguished by the invasion of their linguistic cousins, the Five Nations of Iroquois (later Six Nations), but the Erie were less fortunate. These Iroquoian speaking Indians occupied the south shore of the Great Lake that bears their name, but disappeared from history by 1656, most of them being slain by the Iroquois, although some survivors were adopted by the Iroquois and other tribes.

According to Father Francois Le Mercier (1654), the Erie were called "the Cat Nation, because there is in their country a vast quantity of wild Cats." As earlier shown, some writers believed that the name represented a totemic animal. Erie is an abbreviation of *Eriehronnon* or *Errieronnon* (or *-ronon*), of which the last element, *ronnon,* signifies people or tribe.[27]

Part of Lake Erie lies within the statutory boundaries of Michigan. The name Erie is also given to a village and township in Monroe County, because they border on the lake. These are the only memorials in Michigan to those unfortunate people. However, their name is on places in ten other states and two Canadian provinces.

## The Iroquois

In 1662 an Iroquois war party of one hundred encamped on the shore of Whitefish Bay of Lake Superior, in present Chippewa

County, only a few miles west of Sault Ste. Marie. They were soon discovered by an equally numerous band of Ojibwa, Ottawa, and allied Indians, who made a surprise attack that annihilated the invaders. The Iroquois never again tried to advance into Ojibwa territory.

The victorious Indians ever after called this place *Nadowegoning,* "place of Iroquois bones." It remains Nodoway Point on maps of the United States Geological Survey, while it is Point Iroquois on the Army Corps of Engineers navigational charts. Just offshore is Iroquois Island, and inland from the point is Iroquois Mountain. These toponymic memorials to an Indian battle witnessed by no whites (though described by them from contemporary Indian accounts) are not the only occurrences of the name Iroquois in Michigan. Lake Iroquois is in Oakland County and Iroquois Park is in Genesee County, but they have no historic significance.[28]

There are several opinions on the meaning of Iroquois. According to Charlevoix (1721),

> The name Iroquois is purely French, and is formed from the term *Hiro,* or *Hero,* which means *I have said*—with which these Indians close all their addresses, as the Latins did of old with their *dixi*—and of *koué,* which is a cry sometimes of sadness, when it is prolonged, and sometimes of joy, when it is pronounced shorter. Their proper name is Agonnonsionni, which means cabin-makers, because they build them much more solid than the other Indians.[29]

From Charlevoix's own words, it is clear that the name Iroquois is not "purely French." Francis Parkman held that his view "may be a conversion of the true name with an erroneous rendering." Cadwallader Colden wrote (1727) that the Iroquois "call themselves Onqua-honwe; that is, Men surpassing all others." John Swanton believed that the name was from Algonquin *irinakhoiw,* "real adders," with the addition of the French suffix *-ois.* Beauchamp held (1907) that the term Iroquois was in use before the arrival of the French and suggested that it originated in Iroquois words meaning "to smoke" or "bear."[30] The "bear" explanation was given to me (1956) by William Skenandore, a member of the Oneida tribe of Wisconsin. "Bear" is *hkwari* in Mohawk and *nyak-*

*wai* in Seneca.[31] It is possible that Iroquois could have evolved from those terms, with a French ending added.

The Iroquois's own name, however, was and is *Hodenosaunee,* "people of the longhouse," referring metaphorically to their confederacy. According to Lewis H. Morgan,

> it grew out of the circumstance that they likened their confederacy to a long house, having partitions and separate fires, after their ancient method of building houses, within which the several nations were sheltered under a common roof. Among themselves they never had any other name. The various names given to them at different periods were entirely incidental, none of them being designations by which they ever recognized themselves.[32]

## Tribes of the Iroquois Confederacy

The names of five member tribes of the Iroquois or Six Nations are on the current map of Michigan, and the sixth was there once, although none of these tribes lived in the state. Usually the presence of these names is a record of settlements made by white immigrants from New York state, where places named for the indigenous Iroquois tribes are common.

Cayuga was the name of a post office in Jackson County from 1839 to 1861, having the name of the Iroquois tribe that was located between the Seneca and Onondaga in New York's Mohawk Valley, where their name is on Cayuga Lake and Cayuga County. Details on the reason for the name in Michigan are not at hand, but it is a safe guess that it was transferred to Michigan by settlers from New York state. The meaning of Cayuga, according to Morgan, is "mucky land."[33]

Mohawk, in Keweenaw County, is named for the Mohawk Mining Company, which began copper mining operations at that place in 1898. There is also a Mohawk Lake in Oakland County. The name is that of the Iroquois tribe that lived near Schenectady and Albany, New York, and were called "keepers of the eastern door" of the Long House. Mohawk is of New England Algonquian origin and means "man eaters." Their name for themselves, *Kaniengehaga,* means "people of the place of flint."[34]

Oneida, in Eaton County, was first settled by people from New York state. Another Oneida, in Lenawee County, was named for Oneida, New York, from whence many of the early settlers came. Neither of the Michigan towns survived, but their name continues on Oneida Township in Eaton County. There is also an Oneida Lake in Livingston County and a community called Port Oneida in Leelanau County. The city of Oneida, Oneida Lake, and Oneida County in New York perpetuate the name of the Iroquois tribe, which once possessed the surrounding territory. Most of their descendants live today near Green Bay, Wisconsin, although there are fragments with the Onondaga at Syracuse and on the Grand River, in Ontario. Their name has been translated as "standing stone," for a monument that was their national emblem. Others translate the name into "granite people" and "stone people."[35]

The village of Onondaga, in Ingham County, is named for its township, which was named by Orange Phelps for his old home in Onondaga County, New York. The Onondaga, being at the center of the Long House, were the keepers of the national fire, which meant that their main village was the capital of the confederacy. They still occupy a reservation at the edge of Syracuse, New York. The name Onondaga is interpreted to mean "on the hill."[36]

The village and township of Seneca, in Lenawee County, are named for Seneca County, New York, the former home of many of their early settlers. There is also a Seneca Lake in Keweenaw County. The Seneca tribe for which these places were ultimately named were the "keepers of the western door" for the Long House and still occupy three reservations in western New York. Seneca is a Europeanized corruption of an Algonquian name, which in this form is identical to that of a Roman philosopher. It is from a Mohican or Munsee term signifying "place of the stone," or perhaps "stony earth," from *assineka.* Essentially it is the same as the name of Ossining, New York, in former Mohican territory, which whites made into Sing Sing. Seneca is a popular place-name, found today in fourteen states.

The Seneca of course did not recognize this foreign name; their own name was *Nundawaona*, "great hill people." That was shortened to Nunda for a town in Livingston County, New York, from which it was transferred to townships in Cheboygan County, Michigan, and Lake County, Illinois.[37]

Tuscarora Township in Cheboygan County bears the name of

the last tribe that joined the Six Nations, following their expulsion from North Carolina, which commenced in 1713. Their present reservation is in Niagara County, New York. Their name may have been brought to Michigan from the town of Tuscarora, in Livingston County, New York, from which county many early Michigan settlers came. The name has been interpreted in two ways: "hemp gatherers" and "shirt wearers."[38]

## Other Tribes

A few other Michigan places are named for Indian tribes that not only had no connection with Michigan, but also had no known relation to any of its inhabitants. For some of them details are not available, but these names appear to represent nothing more than the romantic interest in Indians of the "Wild West" that was current in the last century. One of these is Cheyenne Point, on the southwest corner of Beaver Island, in Lake Michigan, Charlevoix County. The Cheyenne, who now live in Montana and Oklahoma, are a western Algonquian tribe that participated with the Sioux in the battle against Custer on June 25, 1876. Their name is held to mean "people of alien speech."[39]

The name of Gomanche Creek in Baraga County undoubtedly was intended by some white name giver to commemorate the Comanche tribe that occupied west Texas before being resettled in Oklahoma. These Shoshonean speakers did not choose the name by which they are known, but received it ultimately from the Utes via the Spanish. Its original meaning was "enemies."[40]

Kickapoo Lake in Gogebic County is named for an Algonquian tribe that was first found by white explorers in Wisconsin. Most of them were in Illinois until the 1830s, when they were removed first to Missouri, then to Kansas, and finally to Oklahoma. One band fled to Mexico and remains there, while one group also remains in Kansas. Their tribal name has been connected to *Kiwigipawa,* meaning "he stands about" or "he moves about, standing now here, now there." Schoolcraft, however, thought the tribal name was taken from a term meaning "otter's ghost."[41]

Lacota, in Van Buren County, has a name that is the Teton dialect version of Dakota, the name for a vast alliance of Plains tribes. The Santee dialect name, Dakota, is attached to a creek in Baraga County. To most people, these Indians are better known by their Ojibwa-French name, Sioux (from *Nadowessioux*), from which comes the name of their language stock, Siouan. The name Dakota or Lakota means "allies." The Michigan place, however, was not named directly for the tribe. The name was chosen by V. D. Dilley, whose father was reading a novel in which the chief character was an Indian maiden named Lacota. The unknown author inappropriately made a personal name out of it.[42]

Mandan, in Keweenaw County, flourished briefly as a copper-mining center in the early years of this century. It was named for a sedentary tribe of Siouan speech that today lives in their old home on the upper Missouri, in North Dakota, with their associated tribes, Arikara and Hidatsa. They were noted for successfully growing several varieties of corn at their northern location. The name Mandan is from the name *Mawatadan* given to them by the Dakotas, but its meaning is not known.[43]

Mohican Lake in Livingston County perpetuates the name of an Algonquian tribe whose name was made prominent by James Fenimore Cooper's novel *The Last of the Mohicans.* The Mohicans occupied the upper Hudson Valley and the western parts of Massachusetts and Connecticut. They were culturally and linguistically, but not politically, identical with the Mohegans of eastern Connecticut. Surviving Mohicans today live in scattered fragments, one of which is known as Stockbridge (from their former residence about Stockbridge, Massachusetts). They share a reservation with the Munsees (Muncies) in Shawano County, Wisconsin. The name Mohican means "wolf."[44]

Muncie Lake, in Grand Traverse County, has the name of a subtribe of the Delawares. No part of the Delawares was ever recorded as resident in Michigan, except for Indian settlers at a Moravian mission called Gnadenhütten in Macomb County, 1782–86. It was named for a similar settlement in Ohio that was wiped out by Kentucky militia during the Revolutionary War. The Muncie lived in eastern Pennsylvania from the time of the earliest white settlement until they began to move into Ontario, Ohio, and Indiana in the late eighteenth century. A band of them living in New York was eventually removed to Wisconsin, while the Muncie in Indiana were removed to Kansas and finally Oklahoma.

The name Muncie appears in the literature as Minsi, Monsey, Muncy, Munsee, and in other ways. The Reverend John Heckewelder derived their name from the Delaware term for "wolf," the name of a clan. Most writers, however, link the name with native words signifying "people of the stony country," or some other phrase involving stones.[45]

The Shawnee did not live in Michigan in historic times, although the name of one of their notable leaders, Tecumseh, is on a town in Lenawee County. There is no historic reason for the presence of the names of Shawnee Lake in Chippewa County and Shawnee Park in Kent County. The Shawnee lived as far south as Georgia before they appeared in Pennsylvania and Ohio in the eighteenth century, although at that time some still lived in amity with the Creeks in Tennessee. Modern Shawnee live in Oklahoma. The name Shawnee comes from *shawunogi,* "southerners."[46]

Tonkawa Lake in Marquette County is far out of place in Michigan. The home of the Tonkawa tribe for which it is named was in central Texas, and their tribe constituted an independent language stock. Their name, meaning "they all stay together," was given by the Caddoan-speaking Waco. A handful of Tonkawa still lives in Oklahoma.[47]

Waco Lake in Delta County is named for an obscure branch of the Tawakoni tribe of Texas. They are mainly remembered because they lived on the site of the present city of Waco, Texas, which took their name. Places in nine other states adopted it, although there is no satisfactory explanation of its meaning.[48]

The village of Yuma in Wexford County is named for a tribe of the lower Colorado River valley of Arizona and California. The name is said to be derived from a titular name meaning "son of the Captain," which was misunderstood as a tribal name by the Spanish. They call themselves Kwichan, sometimes written as Quechuan. Today Yuma is an Arizona city on the lower Colorado River. The reason for the adoption of the name in Michigan is not known.[49]

Perhaps Henry R. Schoolcraft provided an explanation for the retention of many of these names. "The sonorousness and appropriate character of the Indian names," he wrote, "has often been admired. They cast, as it were, a species of drapery over our geography."[50]

## Notes

1. Diamond Jeness, *The Indians of Canada,* 7th ed. (Toronto: University of Toronto Press, 1977), 274–76.

2. Henry R. Schoolcraft, *Information Respecting the History, Condition and Prospects of the Indian Tribes of the United States* (Philadelphia: Lippincott, Grambo, 1851–56), 1:206; 2:358; William W. Warren, *History of the Ojibway Nation* (reprint, Minneapolis: Ross & Haines, 1970), 33; Hodge, *Handbook of American Indians* 1:28.
3. Schoolcraft, *Indian Tribes,* 6:483, 5:40.
4. William Keating, *Narrative of an Expedition to the Sources of St. Peter's River. . . .* (reprint, Minneapolis: Ross & Haines, 1959), 2:147.
5. John R. Swanton, *The Indian Tribes of North America,* Bureau of American Ethnology Bulletin no. 145 (Washington, D.C.: U.S. Government Printing Office, 1952), 260; Warren, *Ojibway Nation,* 36, 82.
6. Baraga, *Otchipwe Language* 1:300.
7. George Lemoine, *Dictionnaire Francais-Montagnais* (Boston: W. B. Cabot & P. Cabot, 1901), 280.
8. Warren, *Ojibway Nation,* 31, 82; see also Hodge, *Handbook of American Indians* 2:167; compare *atâwe,* "trade," in Baraga, *Otchipwe Language* 2:54.
9. Otho Winger, *The Potawatomi Indians* (Elgin, Ill.: Elgin Press, 1939), 147–49; R. David Edmunds, *The Potawatomis, Keepers of the Fire* (Norman: University of Oklahoma Press, 1978), 273–75.
10. Warren, *Ojibway Nation,* 32, 82.
11. Keating, *Narrative of an Expedition* 1:88; see also Virgil J. Vogel, *Indian Place Names in Illinois* (Springfield: Illinois State Historical Society, 1963), 115–16.
12. Gagnieur, "Indian Place Names," (1918), 552; George R. Fox, "Place Names of Berrien County," *Michigan Historical Magazine* 8 (January, 1924): 9.
13. Kellogg, *Early Narratives,* 344; Jonathan Carver, *Travels Through the Interior Parts of North America* (reprint, Minneapolis: Ross & Haines, 1956), 29.
14. Herbert W. Kuhm, "Indian Place-Names in Wisconsin," *Wisconsin Archeologist,* n.s., 33 (March & June, 1952): 88, citing Nicolet and Shea; Hodge, *Handbook of American Indians* 2:82–83; Swanton, *Indian Tribes,* 243–44; Chrysostom Verwyst, "Geographical Names in Wisconsin, Minnesota, and Michigan Having a Chippewa Origin," *Collections State Historical Society of Wisconsin* 12 (1892): 395.
15. Kellogg, *Early Narratives,* 158, 230–31.
16. Albert E. Jenks, "The Wild Rice Gatherers of the Upper Lakes," in *Nineteenth Annual Report, Bureau of American Ethnology,* Part 2 (Washington, D.C.: U.S. Government Printing Office, 1901).
17. Warren, *Ojibway Nation,* 33; Chief Simon Pokagon, *Ogimawkwe Mitigwaki (Queen of the Woods)* (Hartford, Mich.: C. H. Engle, 1899), 124; Walter J. Hoffman, "The Menomini Indians," *Fourteenth Annual Report, Bureau of American Ethnology,* Part 1 (Washington, D.C.: U.S. Government Printing Office, 1896): 12–14; Hodge, *Handbook of American Indians* 1:842.
18. Swanton, *Indian Tribes,* 237–39; C. C. Trowbridge, "Meearmeear Traditions," *Occasional Contributions, University of Michigan Museum of Anthropology* 7 (1938): 1–191.
19. Hodge, *Handbook of American Indians* 2:480; Keating, *Narrative of an Expedition* 1:223–24; Thomas Forsyth, "An account of the Manners and Customs of the Sauk and Fox Nations of Indians Tradition," in Emma H. Blair, ed., *The*

*Indian Tribes of the Upper Mississippi and Region of the Great Lakes* (Cleveland: Arthur H. Clark Co., 1911; reprint, New York: Kraus Reprint Co., 1969), 2:183.

20. Thwaites, *Jesuit Relations* 8:177, 294, 307.
21. Thwaites, *Jesuit Relations* 16:229–31; 38:249; Pierre Francois Xavier de Charlevoix, *History and General Description of New France,* ed. John G. Shea (New York: John G. Shea, 1872), 2:71; Hodge, *Handbook of American Indians* 1:584.
22. Russell Errett, "Indian Geographical Names, Part II," *Magazine of Western History* 2 (July, 1885): 238–40; Lewis H. Morgan, *League of the Ho-de-no-sau-nee or Iroquois* (reprint, New Haven: Human Relations Area Files, 1954), 1:39n.; William M. Beauchamp, *Aboriginal Place Names of New York* (Albany: New York State Museum, 1907; reprint, Detroit: Grand River Books, 1971), 242.
23. Kellogg, *Early Narratives,* 54–55; Louis Hennepin, *A Description of Louisiana* (New York: John G. Shea, 1880; reprint, Readex Microfilm, n.d.), 69–70.
24. Thwaites, *Jesuit Relations* 16:227.
25. Louise P. Kellogg, *The French Regime in Wisconsin and the Northwest* (Madison: State Historical Society of Wisconsin, 1925), 56; Elisabeth Tooker, *An Ethnography of the Huron Indians, 1615–1649,* Bureau of American Ethnology Bulletin no. 190 (Washington, D.C.: U.S. Government Printing Office, 1964), 9 (Tooker's brackets); Swanton, *Indian Tribes,* 233; Jacob P. Dunn, *True Indian Stories, with Glossary of Indiana Indian Names* (Indianapolis: Sentinel Publishing Co., 1909; reprint, North Manchester, Ind.: Lawrence A. Schultz, 1964), 319.
26. Winger, *Potawatomi Indians,* 85; Federal Writers Program (FWP), *Michigan: A Guide to the Wolverine State* (New York: Oxford University Press, 1956), 397; Romig, *Michigan Place Names,* 408; Charles J. Kappler, ed., *Indian Treaties* (reprint, New York: Interland Publishing Co., 1972), 871; A. D. P. Van Buren, "Indian Reminiscences of Calhoun and Kalamazoo Counties," *Michigan Pioneer and Historical Society Collections* (MPHSC) 10 (1886): 155; also statement in annual minutes, June 13, 1889, MPHSC 14 (1889): 28; Lewis Cass, *Remarks on the Condition, Character, and Languages of the North American Indians* (Boston: Cummings, Hilliard & Co., 1826), 28 (extract from *North American Review* 50 [January, 1826]).
27. Thwaites, *Jesuit Relations* 41:10; see also 21:313–15, and Charlevoix, *History of New France* 2:2, 266; Baron Louis Armand de Lahontan, *New Voyages to North America,* ed. R. G. Thwaites (Chicago: A. C. McClurg Co., 1905), 1:320, 321n.
28. Memoir of Nicolas Perrot in Blair, *Indian Tribes* 1:178–81; Kellogg, *French Regime,* 117; Thwaites, *Jesuit Relations* 48:75–79; Warren, *Ojibway Nation,* 147; Schoolcraft, *Indian Tribes* 4:383.
29. Charlevoix, *History of New France* 2:189.
30. Francis Parkman, *The Jesuits in North America in the Seventeenth Century* (Boston: Little Brown, 1896), xlvii, xlviiin.; Cadwallader Colden, *The History of the Five Indian Nations* (New York: New Amsterdam Book Co., 1902), 1:xvii; Beauchamp, *Aboriginal Place Names,* 191, citing Hale.

31. Gunther Michelson, *A Thousand Words of Mohawk* (Ottawa: National Museum of Man, 1973), 137; Wallace Chafe, *Seneca Morphology and Dictionary* (Washington, D.C.: Smithsonian Institution Press, 1967), 74.
32. Morgan, *League of the Iroquois* 1:48.
33. Ibid. 2:129.
34. Romig, *Michigan Place Names,* 375; Hodge, *Handbook of American Indians* 1:921.
35. Hodge, *Handbook of American Indians* 2:123; Beauchamp, *Aboriginal Place Names,* 139–40; Morgan, *League on the Iroquois* 1:49; Henry R. Schoolcraft, *Notes on the Iroquois* (Albany: Erastus H. Pease, 1847), 47, 210.
36. Romig, *Michigan Place Names,* 416; Hodge, *Handbook of American Indians* 2:129.
37. Romig, *Michigan Place Names,* 506; Hodge, *Handbook of American Indians* 2:502; Beauchamp, *Aboriginal Place Names,* 205; E. M. Ruttenber, *Footprints of the Red Men,* Proceedings New York State Historical Association, vol. 6 (1906), 27; Morgan, *League of the Iroquois* 1: flyleaf, 48–49; 2:130.
38. Hodge, *Handbook of American Indians* 2:842; Morgan, *League of the Iroquois* 1:50.
39. Swanton, *Indian Tribes,* 278.
40. Ernest Wallace and E. Adamson Hoebel, *The Comanches, Lords of the South Plains* (Norman: University of Oklahoma Press, 1952), 4–5.
41. Hodge, *Handbook of American Indians* 1:650, 684–85; Schoolcraft, *Indian Tribes* 4:256.
42. Swanton, *Indian Tribes,* 280; Romig, *Michigan Place Names,* 310.
43. Romig, *Michigan Place Names,* 346–47; Stephen R. Riggs, *A Dakota-English Dictionary* (Minneapolis: Ross & Haines, 1968), 309.
44. Mooney & Thomas in Hodge, *Handbook of American Indians* 1:786–89.
45. John G. Heckewelder, *History, Manners and Customs of the Indian Nations. . . .* (Philadelphia: Historical Society of Pennsylvania, 1876; reprint, New York: Arno Press, 1981), 52; Dunn, *True Indian Stories,* 85.
46. Hodge, *Handbook of American Indians* 2:530–36.
47. Ibid., 778–82.
48. Swanton, *Indian Tribes,* 304–5.
49. Hodge, *Handbook of American Indians* 2:1010.
50. Schoolcraft, *Indian Tribes* 5:621.

# III

# Ojibwa Personal Names

Indian personal names were usually given not at birth, but only after some incident or a characteristic of the individual suggested a suitable name. In aboriginal times there were no surnames. Moreover, a name might be changed in the course of one's life in accordance with some circumstance related to dreams, omens, or the exploits of the individual.[1]

Michigan has preserved in its place-names the names of perhaps forty-five individual Indians (some are in doubt), not including persons of purely legendary or literary existence. All of these names were apparently placed on the map by whites, for it was not the Indian custom to glorify individuals in this way. A number of the individuals so honored were considered rogues by contemporary whites, but most of them are remembered as being peaceful and docile in the face of white encroachments.

Some Indian personal names are on the map only in translation (Shavehead, White Pigeon), but the aboriginal forms are in treaties or other documents. The vast majority are preserved on the map in their supposed Indian form, but they are often so mangled that their origin and meaning are difficult to trace.

Some Indians well known in their day are not recognized on the map. Walk-in-the-Water, a notable Wyandot, had his name given, appropriately, to the first steamboat to ply the Great Lakes, but he received no other notice. And there is no place named for Andrew J. Blackbird, one of the few nineteenth-century Indians to become educated and a functionary in white society. Other names on the map recall obscure local figures or commemorate Indians who became nationally famous. The names of great statesmen and warriors such as Pontiac and Tecumseh are preserved not only on the map, but in history and literature. Other names are so obscure that they survive only in treaties and in local place-names. Only one of the names, Pentoga, honors a woman. One name, Bertrand, is the European name a Potawatomi family acquired by intermarriage.

Some personal names on the map, such as Hiawatha and Leelanau, are of legendary or literary figures and will be treated in another place. There are also in this group several pseudo-Indian names invented by white writers.

Personal names on the current state map represent five tribes, but all but two of them are from the Three Fires (Ojibwa, Ottawa, and Potawatomi). The other two are Tecumseh (Shawnee), and Osceola (Seminole). There is no place named for any individual of the Huron-Wyandot, despite their residence in the state for almost two hundred years. The tribes that had virtually abandoned the state before white settlement began left no personal names. These are the Menominee, Miami, and Sauk.

For convenience, we treat the personal names in tribal groups. Since the names were honorific, the places carrying them may sometimes have no direct connection with the persons commemorated or even with their tribes. Usually, however, the individual lived in the vicinity where his name is now found.

The spellings of these names as they appear on the map frequently differ radically from the spellings as they occur in the records—and they also differ greatly in various records. This is to be expected when we are dealing with names from unwritten languages that were recorded by men of varying degrees of literacy, many of whom were unfamiliar with the native languages and their sound systems.

William Webber, an observer at the Treaty of Saginaw in 1819, recalled,

> the secretaries who acted at the making of the treaty do not appear to have taken much pains to make the spelling of the Indian names correspond with the sound, in fact these names were all written by the secretaries and the Indians touched the pen, and the cross followed the name. To illustrate . . . "Neome" is not found as signed to the treaty, and yet he was one of the principal chiefs, but we find "Reaume" signed to the treaty, doubtless intended for "Neome."

The problem is further illustrated by a letter written by Secretary of War Lewis Cass to Surveyor-General Elijah Hayward, April 13, 1835:

> The Original treaties with the Potawatomies . . . have been examined. It appears that the name printed Muck Rose in the latter is written Muck Kose, and it is no doubt the same with Muck kose, signed to the treaty of 1832. As no name, resembling that written Mau Ke Kose in the treaty of 1832, more nearly than these, is signed to that treaty, the presumption is, that the three names designate the same person.[2]

An attempt is made here to provide the English translations of the Indian names from treaties and documents, the works of other scholars, Indian dictionaries and vocabularies, and, occasionally, informants. Frequently, however, the sources give conflicting interpretations, and the dictionaries cannot solve these puzzles for several reasons: they may be incomplete, the name may be a compound word, and, most often, the name is so badly corrupted that it cannot be analyzed.

Short biographical sketches are warranted, in our opinion, because it is probable that most people are interested not only in the names but in their background. For many individual Indians, however, biographical details are unavailable or are buried in remote sources that cannot be found through the limited resources of one individual.

The little village of Assinins on the shore of Keweenaw Bay in Baraga County is on the edge of the Keweenaw Bay Indian community and was founded in 1843 by Father Frederic Baraga. The name is Chippewa for "pebble" or "small stone." Gagnieur wrote (1918) that it was named "after the old chief of years ago." According to the Baraga County Chamber of Commerce, the village was named in 1890 for Chief Edward Assinins. There can be little doubt that Assinins is the name of a local Ojibwa chief of the L'Anse band of Keweenaw Bay, whose name is found in corrupted form in two treaties. In a treaty signed at Fond du Lac of Lake Superior (now Duluth) on August 7, 1847, we find "Assurcens, 2d warrior, his X mark, Ance." The last word, "Ance," is corrupted from L'Anse ("the point"), the French name of his band, indicating their location. "Assurcens" plainly resulted from efforts of a treaty clerk to record the name Assinins. His name appears again as "Aw-se-neece," a headman of L'Anse, in the Treaty of La Pointe, September 30, 1854.[3]

Lake Kawbawgan in Marquette County is named for a local Ojibwa chief, Charles Kawbawgam (so spelled), who died at Marquette from typhoid fever on December 28, 1902. His death certificate in the county clerk's office gives his age as 103, but his obituary says he was born at Sault Ste. Marie "probably about 1819." It is reported that Kawbawgam was the son of chief Shawano of the Sault band of Ojibwa, and the husband of Charlotte, the daughter of chief Maj-e-gee-zik of the Marquette band.

Sometime before 1847 Kawbawgam came to the Marquette area, where he eventually settled at Presque Isle and spent the rest of his life. In 1849 he met the first white settlers of Marquette, a party of iron miners led by R. J. Graveraet. After the Indians were removed to reservations, Kawbawgam continued to live in a cabin in a public park at Presque Isle, eking out a living by hunting and fishing. In 1892 an over-zealous game warden arrested the old man for netting suckers, but the case was dismissed by a justice. In 1899 he lost his sight, but he remained active with the assistance of his wife.

The spelling of the lake name, Kawbawgan, differs from that in the records, which is Kawbawgam. The first is probably correct, but its meaning has not been found.[4]

In the treaty with the Saginaw Bay band of Ojibwa, signed at Saginaw on September 24, 1819, are the following words:

> For the use of Kawkawiskou, or the Crow, a Chippewa chief, six hundred and forty acres of land, on the east side of Saginaw River, at a place called Manitegow, and to include, in the said six hundred and forty acres, the island opposite to such place.

That island was called Crow Island, from the supposed English translation of the chief's name as given in the treaty. The "signature" of the chief at the bottom of the treaty is written Kishkaukou, differing from the spelling used in the text. It is given as Kish-kau-ko in Tanner's captivity narrative, which identifies him as one of the Indians who captured Tanner as a boy in Kentucky. This name was probably further corrupted by whites to Kawkawlin, which is now the name of the river that discharges into Saginaw Bay just north of Bay City, and also of a village in Bay County near the river's mouth.

There is no *l* sound in Ojibwa, so that letter's presence in this name represents a white effort to reproduce an aboriginal phoneme. The whites also misunderstood the meaning of the chief's name, which is not "crow" but "raven," from the root *kakâgi.* (Crow is *endek.*) It has the appearance of an onomatopoeic word, that is, a word that is intended to imitate the sound made by the named animal.

Others point out that the Indian name for the site of Kawkawlin was Oganconning, or "place of the pike fish" or "pickerel." However, that is an entirely different name and is in no way related to Kawkawlin, which several writers have translated as "pike" or "pickerel." In 1670 Father Allouez mentioned a similar name in Wisconsin, Kakaling, which evolved into present Kaukana.[5]

Konteka Creek in Ontonagon County is named for the chief of an Ojibwa band on Ontonagon River who was visited by Henry R. Schoolcraft on July 11 and 12, 1831. His name does not appear in the treaties and has not been translated. Schoolcraft gave this account of his visit:

> *Kon-te-ka,* the chief, and his band saluted us with several rounds of musketry. . . . Afterwards they crossed to our camp and the usual exchange of ceremonies and civilities took place. In a speech from the chief he complained much of hunger, and presented his band as objects of charitable notice. I explained to him the pacific object of my journey . . . and requested the efficient co-operation of himself and his band in putting a stop to war parties . . . against the Sioux [or in support of the Sauks]. Konteka sent me a fine carp [sucker] in the morning. Afterwards he and another chief came over to visit me. The chief said that his child, who had been very ill, was better, and asked me for some white rice . . . for it, which I gave. I also directed a dish of flour and other provisions to enable him to have a feast.

Schoolcraft reported that Konteka "has a countenance indicative of sense and benevolence." The chief told Schoolcraft that his band contained sixty-four members.[6]

Mujekeewis, "West Wind," (d. 1857) was chief of a small Ojibwa band that lived on Thunder Bay River at or near the present site of Alpena. In the Treaty of Washington, March 28, 1836, when the Ojibwa and Ottawa tribes ceded most of their lands in the

Lower Peninsula, several tracts were reserved for Indian use for the next five years, including "one tract of one thousand acres, to be located by Mujeekewis, on Thunder-bay river." Moreover, "Mujeekiwiss" (variant spelling) was listed among chiefs of the "first class" who were to receive five hundred dollars each. Despite these scarcely veiled bribes, his name does not appear among the signers of the treaty. Whatever the treaty terms, Mujekeewis and his band of about twenty-five Indians lived at Partridge Point on Lake Huron, just below Alpena, until his death, which reportedly occurred in 1857. It is said that the chief asserted, some time prior to that date, that he was 110 years old. A report from 1840 claimed that "He had seen nearly, if not quite, one hundred winters." These claims differ from facts given by McKenney and Schoolcraft, who said that the chief was the son of a chief of the same name who was one of the leaders of the attack on the British occupiers of Fort Michilimackinac on June 2, 1763. The younger chief reported that he was born after that event and that his father died during a treaty council on the Maumee. This was probably the Treaty of Fort Industry, July 4, 1805. A Chief Mujekeewis who sided with the British during the War of 1812 must therefore have been the younger one. His age in 1840 could not have been more than 75, and in 1857, the reported date of his death, he could not have been more than 94.

The name he bore appears to be a variant spelling of *Mudjekeewis,* the name of the mythical West Wind spirit, father of Hiawatha in Longfellow's poem, who embraced and then abandoned Winona. However, in this instance, the name may represent the Ojibwa term for firstborn son, which has been given as *Madjikiwiss* and *Mudjekeewis.* The second of the historic bearers of this name is commemorated, in yet another spelling variant, in the name of Michakeewis Park in the city of Alpena.[7]

Naubinway is the name of a village on the Lake Michigan shore, as well as of an island and reef, in Mackinac County. On the surface the name seems to be a variant spelling of Nabunway, an Ojibwa chief mentioned in a letter by Schoolcraft, February 14, 1838, whose name appears again as "Aubunway, of Mille Coquin on the strait" in a treaty signed at Washington on March 28, 1838, and as Naw-aw-bun-way in a treaty signed at La Pointe, Wisconsin, on September 30, 1854.

The treaties give no translation of this name, but Aubunway, in the first treaty, approximates *Aiabêwaiân* in Baraga's dictionary, which is there equated with *Nâbéwaian* and defined as "skin of a male quadruped." Other approximations of the name, in part, are *Nibi* (summer), *Nibowin* (death), and *Namebin* (sucker). The last involves transposition of the syllable *bin* from the end of the word. Such transpositions in Indian names are not rare.[8]

The Ojibwa branded their Iroquois and Sioux enemies as *Naudowaig,* implying, according to Warren, "our enemies," but literally meaning "like unto adders." From that name came dozens of geographic names in seven states and one Canadian province, variously spelled Nadawah, Nodaway, Nottowa, and Nottoway. Some Algonquians applied this name to tribes of Iroqouis speech even if they were allies instead of enemies. It was therefore used as a name for the Wyandot. The pejorative term was also used, however, as a personal name among the Algonquians. Nottowa Township in Isabella County is reportedly named for a local Ojibwa chief who died in 1881 at the age of one hundred. In the Treaty of Isabella, October 18, 1864, his name is listed as Naw-taw-way. A Potawatomi also bore this name, which is found in Calhoun County and St. Joseph County place-names.[9] (See chap. 5.)

The name of Ogemaw County is the Ojibwa word for "chief." The name in this instance is an abbreviation from the name of Ogema-kegato (1794–1840), a chief of the Saginaw band. The county was named in the year of his death. His name appears as Ogemaunkeketo in a treaty signed at Saginaw, September 24, 1819, and O-ke-mau-ke-keto in a petition of October 3, 1832, asking the Congress to honor Article VI of that treaty and reimburse Indians for improvements on ceded lands. It is written by clerks as Ogisna Kegido, Ogima Keegido, and Ogima Kegito in treaties signed in 1837, 1838, and 1839. In the Treaty of Flint River, December 20, 1837, with Commissioner Henry R. Schoolcraft, the name is translated "The Chief Speaker." Ephraim Williams, an early settler in the Saginaw area, declared, "The old chief speaker, O-Gee-Maw-Ke-Ke-To, was the head chief and business manager of the Saginaw Indians." The chief is buried in Roosevelt Park, Bay City, where a bronze plaque attached to a boulder identifies him as Chief Ogemakegate.

The second part of his name, *ke ka to, kegido,* and so on, is a close match with terms for "speak," "speaker," or "speaking" in Baraga's dictionary: *gigit, gigitowin,* and so on. Ogemaw Creek in

Baraga County is not named for this individual, but apparently from the generic word for chief.[10]

The town of Okemos (Little Chief), a suburb of Lansing in Ingham County, was named in 1862 in honor of a Saginaw band Ojibwa chief (1782?–1858). Local historians say that during the War of 1812 he attacked a United States cavalry unit near Sandusky with a band of pro-British Indians and was severely wounded. After this, he never fought again. According to Emerson Greenman, the chief was a nephew of Pontiac and was second in command to Tecumseh. He writes, "From 1839 to about 1858 Okemos and his band wandered around Michigan between Lansing, Saginaw, and Detroit, trading baskets and other native wares for food."

In a treaty signed at Saginaw on September 24, 1819, his name is listed as Okemans, while in another treaty, signed at Flint on December 20, 1833, it is recorded as "Ogimous (The Little Chief, or chief of subordinate authority)." According to one pioneer memoir, "Okemos died at his camp on the Looking-glass [river] above DeWitt [Clinton County] in the year '58." Another source claims that he was buried December 5, 1858. However, attorney William Webber mentioned court testimony by Okemos at Saginaw in 1860, in which the chief said: "I am 76 years old; have lived in Michigan 48 years."[11]

The city of Owosso (formerly Owasso) in Shiawassee County is reportedly named for the Ojibwa chief Wasso. It is more common for whites to drop letters than to add them to Indian names, but this appears to be an exception. An old settler, committing to poetry his information on place-names in the Saginaw valley, wrote these lines:

> From this point we will onward go,
> Till we reach the site of Owosso.
> Eight and fifty years ago,
> As the writer very well doth know,
> This place was reached by two young men,
> They were A. L. Williams and brother Ben,
> Where they found the Indian chief, Wasso,
> And gave it his name with the prefix O.[12]

That individual is listed as a signer of four treaties, if we discount spelling variations. In a treaty signed at Saginaw on September 24, 1819, it is given as Wassau, which means "far off." (Wausau, Wisconsin, has a different origin but the same meaning.) At Washington on March 28, 1828, he was grouped with the "third class" of chiefs, who were to receive $100 each. In a treaty signed at Detroit on January 14, 1837, he is merely listed. In his last treaty, signed at Flint River on December 20, 1837, his name is interpreted as "The Bright Light, or light falling on a distant object." In all but the first treaty, his name is spelled Wasso. The name of an Ottawa Indian, Waus-so, was interpreted as "Lightning" by John Tanner.[13]

Henry R. Schoolcraft translated the name Owasso as "glittering water." He mentioned an earlier individual, whose name he spelled Owassa, who was a chief of the "Saulteurs" (Ojibwa) at Saginaw during Pontiac's rebellion (1763). That chief, upon learning that his nephew had been slain, captured Captain Donald Campbell and had him executed. Actually, that chief's name was Wasson, according to all accounts. This became Wauseon as the name of a city in Fulton County, Ohio, where it is interpreted as "far off," the same as Wausau, Wisconsin. Owosso is named either for Schoolcraft's Owasso or for the later chief. Since so few Indian names from the presettlement period survive as place-names, it is probable that the city is named for the Wasso whose name is in the treaties.

Names similar to Owosso exist in several states: Owasso Lake in New Jersey, Owasa in Iowa, Owasso in Alabama and Oklahoma. Some of these names could have been derived from Schoolcraft's legend of "Owasso and Waywoond" in *Algic Researches* or corrupted from *Owaissa,* "the bluebird" in Longfellow's *Hiawatha.*[14]

Pentoga is perhaps the only place in Michigan that is named for an Indian woman, aside from the legendary and literary figures mentioned elsewhere. It is the name of a village and county park in Iron County, honoring Pentoga Edwards, wife of a local Ojibwa chief. Chief Edwards, leader of an Indian band at Chicagon Lake in the late nineteenth and early twentieth centuries, is commemorated by Chief Edwards Lake in Iron County.[15]

The Peshekee ("Buffalo") River is a tributary of Lake Michigamme in Marquette County. It is named for an Ojibwa chief of La Pointe (Chequamegon), Wisconsin, who signed treaties in 1826 and 1837. His remarks at the Treaty of Fond du Lac (Duluth), signed August 5, 1826, were recorded by Thomas L. McKenney in *Sketches of a Tour to the Lakes.* His name in the treaty is spelled Peexhickee. The *x* is undoubtedly a mistake, for the name of another chief from River St. Croix is written Peezhickee. In a treaty signed with Governor Henry Dodge of Wisconsin Territory at St. Peter's, Minnesota (now Mendota) on July 29, 1837, Chief "Pe-zhe-ke or the Buffalo" is listed as a signer from La Pointe. An earlier chief of the same name, "Peshawkay or Young Ox," signed the Treaty of Greenville, Ohio, in 1795. Although this name is variously spelled in documents, it is always translated the same way, "Buffalo," or its equivalents: Pisikious (*boeufs sauvages,* "wild cattle") by Marquette (1673), Peshekey by John Long (1796), Pe-zhe-ke by John Tanner (1830), and Pijiki by Father Baraga (1881).[16]

Pewamo, a village in Ionia County, was named in 1859, according to Romig, at the suggestion of "J. C. Blanchard, for an Indian chief with whom he had hunted along the Grand River." Henry Gannett (1905) called Pewamo the son of Shacoe, an Ojibwa chief. Neither of these two names appears in the treaties. Father Verwyst translated the name "From *biwamo* (the trail diverges)."[17]

On its face, the name of Sashabaw Creek in Oakland County appears to be derived from Sassaba, the name of a lesser chief at Sault Ste. Marie in the early nineteenth century. According to Schoolcraft, Sassaba's brother was killed fighting beside Tecumseh on behalf of the British in the battle of the Thames, October 5, 1813. After the war, Sassaba continued to maintain a hostile disposition toward the Americans. In 1820 he still displayed the British flag at his camp, and when General Cass came to the Ojibwa country to arrange a treaty ceding to the United States four square miles at the rapids of St. Mary, Sassaba sent his women and children across the river into Canada. Nevertheless, Cass hauled down the British flag and concluded the treaty with the more amenable chiefs on June 16, 1820. The meaning of Sassaba is undetermined.[18]

The name of the Shebeon River, a tributary of Saginaw Bay in Huron County, is apparently the Ojibwa plural word meaning "rivers, streams," which is given as *sibi-wan* by Baraga. In this instance it probably comes from the name of a Saginaw Ojibwa whose name appears as Sepewan in the list of signers of the Treaty of Saginaw, September 24, 1819.[19]

A cluster of Tawas names in Iosco County includes those of the towns of Tawas City and East Tawas, as well as Tawas Bay,

Lake, Point, and Tawas Point State Park. There is general agreement that the name Tawas is shortened from Ottawas. However, the names of these places are apparently not given for the tribe, as some have supposed, but for the name of a Saginaw band Ojibwa chief. The *s* ending does not make it a plural, as in English, but a diminutive—i.e., Little Ottawa (Little Trader). In a memoir written in 1884, former Indian trader Ephraim S. Williams of Flint recalled these anecdotes of Chief Tawas:

> O-taw-wars [*sic*] had an American officers' uniform, coat, sword, and belt and epaulets, which he obtained in the war of 1812 and kept very choice, as he would show them only when asked. As traders we thought very much of the old chief and his family. He was very intelligent, asking many questions about the doings of the white people. He wanted me to bring him newspapers and read the news to him and explain about the doings of Congress, and the Great Father, the President of the United States. . . .

The chief's name is recorded as "Ottawonce" in a treaty signed at St. Louis, August 24, 1816; as "Ottawaus (The Little Ottawa)" in a treaty signed at Flint River on December 20, 1837, and "Ot-taw-ance, chief," in a treaty signed at Detroit, August 22, 1855.[20]

The Waiska River flows into Waiska Bay of Lake Superior in Chippewa County, just east of the Bay Mills Indian Reservation. Waiska is the spelling of the United States Geological Survey, but the state highway sign (1974) spelled it Waiskee, which more nearly approximates the name as it appears in treaties.

Waiskee was a chief of the St. Marys band of Ojibwa whose members once lived at Sault Ste. Marie. Most of them now live at the Bay Mills reservation about twenty miles west of their original home. The chief's name is recorded as Wayishkey in the list of signers of a treaty at the Sault on June 16, 1826, and as Waishkee in a treaty signed at Fond du Lac on August 5, 1826. In the latter treaty one section of land was assigned to two of his children, Waybossinoqua and John J. Wayishkee. His name again appears as Wayishkee in the Treaty of Butte des Morts, August 11, 1827, and "Waish-key, headman" in the Treaty of La Pointe, September 30,

1854. There he is listed under the Mississippi bands, which could be a clerical error or the name of another person.

The word "headman" after Waish-key in the last treaty appears to be not merely a title, but a translation of the name. Father Baraga's *Otchipwe Language* lists *Waiéshkat* as a term meaning "In the beginning, at first." Schoolcraft, apparently in error, said that the name of chief "Wayishkee" meant "Firstborn son."[21] (But compare *Michakeewis* herein; see also *madjikiwiss,* "The firstborn boy of a family," in Baraga, *Otchipwe Language* 1:33.)

Wawatam Township in Emmet County has one of the oldest aboriginal personal names in Michigan. William Warren, Ojibwa historian, wrote that during the Indian capture of Mackinac in Pontiac's rebellion, June 2, 1763, the English trader Alexander Henry "was eventually saved by Wa-wa-tam, or Wow-yat-ton (Whirling Eddy), his adopted Ojibway brother. . . ."

Following Henry's capture, Wawatam first made a speech to the Indians claiming Henry as his adopted brother and offering presents for his release, which was granted. Later, when Indians from Detroit presented a possible threat to Henry's life, Wawatam led him to a safe refuge. According to Henry:

> Wawatam was not slow to exert himself for my preservation; but, leaving Michilimackinac in the night, transported himself and all his lodge to Point St. Ignace, on the opposite side of the strait. Here we remained till daylight, and then went into the Bay of Boutchitaowy [now Batchawana] in which we spend [*sic*] three days in fishing and hunting, and where we found plenty of wild fowl. Leaving the bay we made for Isle aux Outardes, where we were obliged to put in on account of the wind's coming ahead.

Inquiring about the ultimate fate of Wawatam, Henry R. Schoolcraft was informed in November, 1833, by Mrs. Michael Dousman of the trading family that "Wawetum became blind, and was burned, accidentally, in his lodge at the point (Ottawa Point)." The date of this incident was not given.[22]

Wawatam was commemorated not only in a township name. The railroad car ferry that formerly crossed the Straits of Mackinac was called Wawatam.

## Notes

1. For an introduction to Indian naming customs, see John R. Swanton, "Names and Naming," in Hodge, *Handbook of American Indians* 2:16–18.
2. William Webber, "Indian Cession of 1819, Made by the Treaty of Saginaw," *MPHSC* 26 (1894–95): 517–34; Clarence E. Carter, ed., *Michigan, 1829–37,* The Territorial Papers of the United States, vol. 12 (Washington D.C.: U.S. Government Printing Office, 1934–58), 895–96.
3. Gagnieur, "Place Names" (1918), 340; Romig, *Michigan Place Names,* 32; Kappler, *Indian Treaties,* 569, 651.
4. Copy of record of death No. 147–564, certified by Henry R. Skewis, clerk of Marquette County, November 8, 1978; clippings dated only 1902 from the *Marquette Mining Journal;* Ralph D. Williams, *Honorable Peter White* (n.p.: Penton Publishing Co., n.d.), 25–27; all sources furnished courtesy of Marquette County Historical Society.
5. Kappler, *Indian Treaties,* 186; John Tanner, *Narrative of the Captivity and Adventures of John Tanner* (reprint, Minneapolis: Ross & Haines, 1965), 4–5; Baraga, *Otchipwe Language* 1:63, 206; Romig, *Michigan Place Names,* 298; FWP, *Michigan,* 489; Elijah Haines, *The American Indian* (reprint, Evansville, Ind.: Unigraphic, 1977), 731; still another writer gave the early name of the river as O-kaw-kaw-ning but did not explain it. Ephraim Williams, "Personal Reminiscences," *MPHSC* 8 (1885): 251. On the Wisconsin name see Kellogg, *Early Narratives,* 150.
6. Henry R. Schoolcraft, *Personal Memoirs of a Residence of Thirty Years with the Indian Tribes. . . .* (reprint, New York: Arno Press, 1975), 359–60.
7. Kappler, *Indian Treaties,* 77–78, 451, 455; David D. Oliver, *Centennial History of Alpena County, Michigan, from 1837 to 1876* (Alpena: Argus Printing House, 1903), 48; David A. Armour, ed., *Attack at Michilimackinac, 1763* (Mackinac Island: Mackinac Island State Park Commission, 1971), 103n.; Schoolcraft, *Personal Memoirs,* 447; Thomas L. McKenney, *Sketches of a Tour to the Lakes* (Barre, Mass.: Imprint Society, 1972), 408, 410; Baraga, *Otchipwe Language* 2:203.
8. Kappler, *Indian Treaties,* 69; Chase S. Osborn and Stellanova Osborn, *Schoolcraft, Longfellow, Hiawatha* (Lancaster, Pa.: Jacques Cattell Press, 1942), 586; Baraga, *Otchipwe Language* 1:242, 249, 250, 285.
9. Warren, *Ojibway Nation,* 83; compare Schoolcraft, *Indian Tribes* 5:36; Cass, *Remarks on the North American Indians,* 24; compare Tanner, *Narrative of Captivity,* 71; Romig, *Michigan Place Names,* 407–8; Kappler, *Indian Treaties,* 871.
10. Carter, *Michigan, 1829–37,* William L. Jenks, "History and Meaning of the County Names of Michigan," *MPHSC* 38 (1912): 468; Williams, "Personal Reminiscences," 255; Baraga, *Otchipwe Language* 1:239; 2:131; FWP, *Michigan,* 203.
11. M. A. Leeson, *History of Kent County, Michigan* (Chicago: C. C. Chapman Co., 1881), 56–57, 156; Greenman, *Indians of Michigan,* 38; Kappler, *Indian Treaties,* 187, 502; the definition "little chief" is also given in Verwyst,

"Geographical Names," 395; Mrs. M. J. Niles, "Sketch of Old Times in Clinton County," *MPHSC* 14 (1889): 624; Webber, "Indian Cession of 1819," 523.

12. Albert Miller, "The Rivers of the Saginaw Valley Some Sixty Years Ago," *MPHSC* 14 (1899): 505.
13. Romig, *Michigan Place Names,* 424; Kappler, *Indian Treaties,* 187, 454, 485, 502; Tanner, *Narrative of Captivity,* 35; D. G. Emmert, "The Indians of Shiawassee County," Pt. 2, *Michigan History* 47 (September, 1963): 243–72.
14. Schoolcraft, *Indian Tribes* 2:294; 3:509; Francis Parkman, *The Conspiracy of Pontiac* (Boston: Little Brown, 1922), 1:308–10; Howard H. Peckham, *Pontiac and the Indian Uprising* (Chicago: University of Chicago Press, 1961), 194–95; Maria Ewing Martin, "Origin of Ohio Place Names," *Ohio Archaeological and Historical Society Publications* 14 (1905): 288; Mentor Williams, *Schoolcraft's Indian Legends* (East Lansing: Michigan State University Press, 1956), 215–21.
15. Letter from Harold Bernhardt, president of Iron County Historical and Museum Society, January 20, 1984.
16. McKenney, *Sketches,* 379–80; Kappler, *Indian Treaties,* 44, 270–71, 493; Kellogg, *Early Narratives,* 237; Long, in Reuben G. Thwaites, ed., *Early Western Travels* (Cleveland: Arthur H. Clark, 1904–7), 2:233 *ff.*; Tanner, *Narrative of Captivity,* 301; Baraga, *Otchipwe Language* 2:354.
17. Romig, *Michigan Place Names,* 439; Henry Gannett, *American Names* (reprint, Washington, D.C.: Public Affairs Press, 1947), 244; Verwyst, "Geographical Names," 396.
18. Schoolcraft, *Indian Tribes* 1:112; 6:386; Kappler, *Indian Treaties,* 187–88.
19. Baraga, *Otchipwe Language* 2:366; Kappler, *Indian Treaties,* 187.
20. Romig, *Michigan Place Names,* 550; FWP, *Michigan,* 467; Verwyst, "Geographical Names," 397; Ephraim Williams, "What I Know About O-taw-wars and Ne-war-Go," *MPHSC* 7 (1881): 137; Kappler, *Indian Treaties,* 133, 502, 735.
21. Kappler, *Indian Treaties,* 188, 270, 273, 282, 652; Baraga, *Otchipwe Language* 2:396; Schoolcraft, *Personal Memoirs,* 290, 303.
22. Warren, *Ojibway Nation,* 206 *ff.*; Armour, *Attack at Michilimackinac,* 99; Schoolcraft, *Personal Memoirs*, 452.

IV

# Ottawa Personal Names

Cobmoosa Lake in Oceana County continues a name that was originally given to a post office from 1878 to 1916. It commemorates Ottawa subchief Cobmoosa (1768–1872), who lived near the lake in his last years. Cobmoosa is mentioned in two treaties. In the first, signed at Washington on March 28, 1836, "Cawpemossay or the Walker" is listed as a "Chief of the first class," each member of which was to receive $500. Despite the bribe for assent to huge land cessions, his name does not appear in the list of treaty signers.

By a treaty signed at Detroit, July 31, 1855, which again does not list Cobmoosa among the signers, the Ottawa and Chippewa Indians of Michigan gave up their land claims and tribal organization in return for a reservation and individual allotments plus benefits of $538,400. One year later, the signature of "Caw-ba-mo-say" is listed as the endorser of Senate amendments to that treaty on behalf of the Grand River band of Ottawa and Chippewa Indians, although his name is absent from the original document being amended.

Reportedly Cobmoosa lived in the early 1850s at or near the site of Lowell, in Kent County, where the Flat River joins the Grand. Following the final treaty, about 1,300 Indians were moved into Oceana County during 1857 and 1858, and a log cabin was built for Cobmoosa three miles west of the post office named for him. It is reported that he died near Pentwater in 1872. The treaty translation of his name, "the walker," may be compared with *bemoosed,* "walker," in Baraga's dictionary.[1]

Good Hart, in Emmet County, occupies part of the site of the Ottawa village of L'Arbre Croche, which flourished along the Lake Michigan shore in the eighteenth and early nineteenth centuries. Its present name is a misspelling of the name of Good Heart, or Kaw-me-no-tea, who was a paternal uncle of the notable Andrew J. Blackbird.

Hazy Cloud County Park in Kent County was named by the

county board in July, 1928, at the suggestion of Charles E. Belknap. Hazy Cloud was chief of an Ottawa band on the Thornapple River, a tributary of the Grand. Little is known of this Indian besides the following memoir by Dwight Goss (1905):

> At Thornapple River, or Ada, there was a small band of Indians of whom Ma-ub-bin-na-kiz-hick, or Hazy Cloud, was the Chief. Although of a small stature, he was a man of commanding influence with his tribe. He was on the most friendly terms with the whites, visited Washington, and was one of the leading spirits of the treaty of 1836.

The only name resembling the above in the Washington treaty of March 28, 1836, is Nabun Agheezhig, of Grand River, which appears in the list of signers of the treaty and of a supplementary article. In the treaty itself, "Nabun Egeezhig son of Kewayquabowequah" is named as a chief of the first class who was to receive $500.[2]

The village of Kewadin occupies the site of a former Indian settlement on the shore of Elk Lake, Antrim County, and preserves the name of a local Ottawa chief. A treaty clerk apparently corrupted it into "Key-way-ken-do, headman," in listing the signers of the Treaty of Detroit, July 31, 1855.

John Tanner, in his captivity narrative, mentions Ke-wa-tin, "the north wind," as an elder brother of Wa-me-gon-a-biew, both of them being members of his adopted family during captivity. The name is variously spelled and is fairly common as both a personal and geographic name. A district of the Canadian Arctic is called Keewatin, a lake in Marquette and Baraga counties of Michigan is named Keewayden, and a resort in St. Clair County is called Keewahdin Beach. The name in one form or another also occurs in Ontario, Minnesota, New York, Maine, and Florida. A chief from Crow River, Minnesota, who signed the Treaty of Fond du Lac, August 5, 1826, was named Keewayden. In Longfellow's *Hiawatha,* Keewaydin is the Northwest Wind. In Baraga's Otchipwe dictionary, Kiwédin is "North Wind." Baraga also wrote it Kiwatin, pronounced *Kiwétin.*[3]

Missaukee County has the name of an Ottawa chief who signed treaties at Maumee Bay (Toledo) on August 30, 1831, and February 18, 1833. The name has been extended to Lake Missaukee, Missaukee "Mountain," and Missaukee Park, a village, all within the county.

The chief's name is spelled Me-sau-kee in both treaties that he signed. Father Verwyst believed that the name was a corruption from *missisaging* ("at the large mouth of a river"). A similar place-name in Ontario, Missisauga River, gave a name to a tribe. However, there is no need to presume corruption of the present name, for it is virtually pure Ottawa, composed of *missi,* "great," and *au-kee,* "the world, the earth, land, country, soil."[4]

Newaygo County, containing the town of Newaygo and Newaygo State Park, is named for a relatively obscure Ottawa band chief. It appears that more than one individual bore this name, and one of them, said to be an Ojibwa, spent his life about Green Point (Saginaw) and the Saginaw Bay region. It is this individual, apparently, whose name appears as Nuwagon in a treaty signed at Saginaw on September 24, 1819. In two accounts written by William R. McCormick, both published in 1883, he is called Neh-way-go and in another, published by Ephraim Williams of Flint in 1884, he is called Ne-war-go. All of them relate apocryphal stories of three violent episodes between Newaygo and other Indians, each of which resulted in the death of his opponent. Eventually, he was reported slain by a relative of one of those he had killed. It is possible that this individual is the one who signed a treaty at Detroit on July 31, 1855, which records his name as "Nay-waw-goo, chief, his x mark." If this is the same individual who signed the Treaty of Saginaw thirty-six years earlier, he must have been of advanced age by that time.

However, it is established that another chief Newaygo was an Ottawa, and it is apparently this individual for whom the county is named. Henry R. Schoolcraft reported that at Mackinac Island on August 19, 1833, he was visited by "Ningwegon (or the Wing)," an Ottawa, accompanied by his band of twenty-eight persons. According to Schoolcraft, the chief was then seventy-six years of age and gray-haired. He was described as a friend of the Americans whose father was a native of Detroit, who became acquainted with General Cass during the War of 1812. At a later date he migrated with his father and grandmother to L'Arbre Croche on Lake Michigan. Schoolcraft identified this chief as the one who was granted an annuity under the treaty signed with the Chippewa and Ottawa tribes at Washington on March 28, 1836. The passage reads:

> The Ottawas having consideration for one of their aged chiefs, who is reduced to poverty, and it being known that he was a firm friend of the American government, in that quarter, during the late war, and suffered much in consequence of his sentiments, it is agreed that an annuity of one hundred dollars per annum shall be paid to Ningwegon, or the Wing, during his natural life, in money or goods, as he may choose.

His name, as Negwegon, was given to present Alcona County in 1840 but dropped three years later. It was then transferred to Newaygo County, which was organized in 1851. The name Newaygo is a corruption of the Ojibwa-Ottawa word *ningwigan*, "wing."[5]

Chief Noon Day Lake in Barry County has the name of an Ottawa chief who had a village on the site of Grand Rapids in 1831. He was converted to Christianity by missionary Leonard Slater and moved with him to Prairieville, Barry County.

It is reported that as a warrior for the British, he was present at the burning of Buffalo, New York, in 1813, and while on a visit to Washington in 1836 identified Richard M. Johnson (vice-president, 1837–41) as the slayer of Tecumseh. It is said that Noon Day went to Kansas in 1828 with missionary Isaac McCoy and five others to look for land for transplanted Indians.

In a treaty signed at Maumee Bay on August 30, 1831, Noon Day assented to the surrender of Ottawa lands in Ohio and the removal of the occupants. The treaty provided "that there shall be allowed to Nau-on-qual-que-zhick, (Noon Day) one hundred dollars, out of the surplus fund accruing from the sales of the land herein ceded, in consequence of not owing any debts, and having his land sold, to pay the debts of his brethren." His name in the signers' list is spelled differently: Nau-qua-ga-sheek.

In another treaty surrendering huge tracts of Michigan that was signed at Washington, March 28, 1836, "Nawequa Geezhig or Noon Day" was counted in the "first class" of chiefs, who were to receive five hundred dollars each for their assent to that document. In the next paragraph are listed chiefs of the "second class," who were to receive two hundred dollars each, including "Nawiqua Geezhig of Flat River." Whether this is the same individual engaging in double dipping, or someone else, is unclear.

The chief's name appears for the last time as Naw-we-ge-zhick, among the signers of a treaty at Detroit, August 2, 1855, with the Saginaw band of Chippewas. Perhaps, since the dispossession of his own people with his consent had eliminated the Ottawa land base, he had joined his Chippewa brethren. However, it is reported that he was last seen alive in Hastings, Barry County, on August 18, 1855, only sixteen days after the treaty. He is said to have died nearby a few days later at the age of ninety-eight.[6]

Pabama (or Pebawma) Lake and Little Pebawma Lake in Oceana County are named for Joseph Pabama (1810–70), an Ottawa from the Grand River valley who moved to Oceana County after the treaty of 1855. He was elected a chief in the same year. Pabama was a Catholic convert and longtime treasurer of Elbridge Township.

His name is listed as Pay-baw-me, of the Grand River bands of Chippewa and Ottawa, in the treaty of July 31, 1855. His name occurs again a year later as a signer of an amendment to this treaty. The signature of L. Pay-baw-maw-she in a treaty signed at Isabella Reservation on October 18, 1864, is a similar name, but may be that of a different individual.

Pabama's name is not translated in the treaties. The terms in Baraga's Otchipwe dictionary that most nearly approximate this name are *babâmisse,* "it is flying about (a bird, etc.)," and *babamossé* (*nin,* "I walk about"). There is no significance to the substitution of *b* for *p* in Ojibwa-Ottawa names, since no distinction is made between those letters or sounds in their language.[7]

In 1852 a Catholic mission to the Ottawa Indians was established on the west side of Grand Traverse Bay, in Leelanau County. Ottawa and Chippewa Indians settled there, and their descendants remain there. In 1883 Father Philip Zara named the village Peshaube, for the local chief, and today it is called Peshawbestown (United States Geological Survey map) and Peshabestown (official state highway map, 1982). The chief was the namesake and apparent descendant of that Pe-shau-be described by John Tanner (1830) as "a celebrated war-chief of the Otawwaws." The later Peshawbe lived after the close of the treaty-making period, and so his name is not in the treaties. No interpretation of his name has been found. The first element in it may represent *pe-zhew,* "wildcat." In Menominee, *peswâba* means "dawn."[8]

The city of Petoskey on Little Traverse Bay, Emmet County,

received its name in 1873 with the establishment of a post office and the coming of the railroad. It honors Chief Ignatius Petosky (1787–1884), thus spelled, whose name is on a local marker. He was the son of Antoine Carre, a French Canadian trader, and an Ottawa woman. Petoskey, as it is more frequently written, was the name of an Ottawa family long resident at that location, where they owned land. Ignatius Petoskey fathered ten children. One of them, Francis Petoskey, is mentioned by two Michigan Indian writers, Andrew Blackbird and Simon Pokagon, as having been their fellow student in the academy at Twinsburg, Ohio. The three were the only Indians attending the school at that time (1845–49).

Besides the city, a nearby state park is named Petoskey. The original form of the name is said to have been Petosega or Petosegay. Two explanations of this name are in print. Ella Petoskey, granddaughter of Ignatius, wrote in 1929 that its broad meaning was "Rising Sun," while Verwyst derived the name from *pitoskig,* "between two swamps." Neither explanation can be reconciled with Ojibwa and Ottawa language sources. In Baraga's Otchipwe dictionary, the nearest approximation to this name is *patashkanje,* "long billed curlew," but the name could be a corruption from several other terms.[9]

One Michigan place-name, Pontiac, is a household word because it was also given to an automobile. Pontiac is the chief city of Oakland County, incorporated as a village in 1837, although its post office was established fifteen years later. The name was extended to the township, to Pontiac Creek and Lake, and to a village and state park named from the lake. Outside Michigan, Pontiac is the name of villages or towns in Illinois, Missouri, Rhode Island, and South Carolina, a bay in Seattle, Washington, and a city and county in Quebec.

Pontiac was an Ottawa chief (ca. 1720–1769) who is regarded as one of the outstanding statesmen and war strategists of the American Indian people. His impact on American history was made in the Indian rebellion against the English in the northwest in 1763. Although he accepted the transfer of authority in the northwest from France to Great Britain after the French and Indian War, he soon became dissatisfied, partly over the lesser generosity of the new occupants of the old French domain and their refusal to advance credit in trade. He united most of the tribes west of Pittsburgh and north of the Ohio, as well as some Seneca, into a military alliance. In late May and early June, 1763, they launched a coordinated attack on all British forts in the west, capturing eight of them. Only Fort Pitt, which was relieved by Colonel Bouquet, and Detroit, which withstood a five-month seige, evaded capture by the Indians. Only a message from the French commander at Fort de Chartres on the Mississippi, which the British had not yet occupied, persuaded Pontiac to give up. Following a preliminary peace agreement signed at Ouiatanon (Lafayette, Indiana) the Indians signed a treaty at Detroit on August 17, 1765. Pontiac's character won warm praise from two Englishmen who dealt with him, Robert Rogers, the ranger, and George Croghan, peace emissary.

One result of the Indian rebellion was the royal proclamation of October 7, 1763, which closed to settlement the newly won English empire west of the Appalachians. That, in turn, was one of the grievances leading the English colonists toward the War of Independence. Another result of the Indian war was the stationing of a permanent armed force in America. That, and the taxes levied to support it, caused more grievances that led to the revolutionary war.

In 1769, while on a visit to St. Louis, Pontiac was killed at Cahokia, Illinois, by an Illionis Indian said to have been bribed by an English merchant. The meaning of Pontiac's name is in doubt, but there is an Ottawa tradition referring to him as Obwandiyag, and his name was reportedly pronounced *Bwon-diac* in Ottawa. According to Blackbird, *bon* or *bwon* means "stopping," and *obwon* means "his stopping" or "stopping it" or "stopping him." The meaning of *diac* or *diag* remains unexplained. Another source (Kelton) derived Pontiac from *banitiyak,* "a stick planted in the ground to anchor (stop) a canoe." Finally, Father Gagnieur gave the chief's name as Bwandiag, saying that *bwan,* an appellation given to the Sioux, referred to a spit for roasting or boiling.[10]

Saubee Lake in Eaton County is probably named for Sawba, an Ottawa underchief in Barry County in the 1830s. This assumption is based on the proximity of the lake to the Thornapple River, where the chief lived. It is reported that Sawba was the son of chief Po-mob-a-koo and was killed by whites near Mt. Pleasant shortly after the Civil War. Sawba or Saubee is apparently a corruption of *sibi,* Ojibwa and Ottawa for "river" or "stream."[11]

Wabaningo, a resort village at the outlet of White Lake into Lake Michigan, in Muskegon County, was reputedly named in

1897 for a local Ottawa chief. In the 1830s he headed a band of a half dozen families who lived at that place and planted corn there. Wabaningo is a branch post office of Whitehall. The name is probably derived from the Ojibwa and Ottawa name for the morning star, *Wabananang*. In Longfellow's *Hiawatha*, Wabun-Annung is the "Star of Morning."[12]

The historic Ottawa village of L'Arbre Croche in Emmet County was called "The Crooked Tree" in English and *War-gun-uk-ke-zee* in Ottawa, according to John Tanner. Tanner also said (1830) that the chief of L'Arbre Croche was named *Wah-ka-zee*. His name compares well with the term *wâgâkasi mitig*, "the tree is crooked," in Baraga's Otchipwe dictionary.

Chief Wah-ka-zee's name was spelled *Wa-ke-zoo* by Ottawa historian Andrew J. Blackbird, who was his nephew. Wa-ke-zoo was a brother of Blackbird's father, Black Hawk. According to Blackbird, Wakezoo died in Manitoba. Etta Wilson recorded his first name as Joseph.

According to Mrs. Wilson, Joseph Waukazoo and several Chippewa and Ottawa Indians from Middle Village (now Good Hart) traveled south to Allegan late in 1837. Their purpose was to persuade Congregational missionary George W. Smith, who was already working among Indians of southwestern Michigan, to be their pastor. Following a meeting in January, 1838, an Indian colony was established near Allegan, with a chapel and school. In 1839 the colony moved to Black River, near present Holland. In 1847 Dutch settlers began the settlement on Black Lake (now Macatawa) that became the city of Holland. Within two years Smith and the Indian colony moved to the west shore of Grand Traverse Bay in Leelanau County. A memorial to their former presence on Black River is the resort village of Waukazoo Park, on the north shore of Lake Macatawa.

In their new location, the Reverend Smith founded the village of Waukazooville, naming it for Peter Waukazoo, son of the chief. In 1852 the new settlement was absorbed by Northport, a white town, which preserves the old name on Waukazoo Street.[13]

## Notes

1. Harry L. Spooner, "Indians of Oceana," *Michigan History* 15 (Autumn, 1931): 654–55; Romig, *Michigan Place Names*, 123; Kappler, *Indian Treaties*, 455,

731; Baraga, *Otchipwe Language* 1:280–81; Franklin Everett, *Memorials of the Grand River Valley* (Chicago: Legal News Co., 1878), 279–81.

2. Good Hart: Romig, *Michigan Place Names,* 228; Blackbird, *History of the Ottawa and Chippewa,* 48. Hazy Cloud: file material from Kent County Park Commission furnished by Mary K. Awdey, public relations and recreation planner, January 24, 1984; Dwight Goss, "The Indians of the Grand River Valley," *MPHSC* 30 (1905): 181; Kappler, *Indian Treaties,* 454, 455, 456.
3. FWP, *Michigan,* 521; Romig, *Michigan Place Names,* 302; Kappler, *Indian Treaties,* 271, 730–31; Tanner, *Narrative of Captivity,* 20; Baraga, *Otchipwe Language* 1:181, 209.
4. Romig, *Michigan Place Names,* 374; Kappler, *Indian Treaties,* 339, 394; Verwyst, "Geographical Names," 393; Blackbird, *History of the Ottawa and Chippewa,* 123.
5. Joseph N. Kane, *The American Counties,* 3d ed. (Metuchen, N.J.: Scarecrow Press, 1972), 270; Kappler, *Indian Treaties,* 187, 270, 454, 730; Williams, "O-taw-wars," 137–39; W. R. McCormick, "A Pioneer Incident," *MPHSC* 4 (1883): 376–79; idem, "Indian Stoicism and Courage," in H. R. Page Co., *History of Bay County, Michigan* (Chicago: 1883); William L. Jenks, "History and Meaning of the County Names of Michigan," *Michigan History* 10 (summary) (October, 1926): 647; Schoolcraft, *Personal Memoirs,* 446; Baraga, *Otchipwe Language* 1:290.
6. Kappler, *Indian Treaties,* 339, 450–56, 735; Emerson Greenman, "Indian Chiefs of Michigan," *Michigan History* 23 (Summer, 1939): 228–29; Charles A. Weissert, "The Indians of Barry County and the Work of Leonard Slater, Missionary," *Michigan History* 16, no. 3 (Summer, 1932): 321–33; Van Buren, "Indian Reminiscences," 158, 160–61.
7. Kappler, *Indian Treaties,* 730–31, 871; H. R. Page Co., *History of Manistee, Mason, and Oceana Counties* (Chicago: 1882), 82; Baraga, *Otchipwe Language* 2:61.
8. Romig, *Michigan Place Names,* 428; FWP, *Michigan,* 531; Tanner, *Narrative of Captivity,* 34, 302; Hoffman, "Menomini Indians," 310.
9. Percy H. Andrus, "Markers and Memorials in Michigan," *Michigan History* 15 (Spring, 1931): 211; Blackbird, *History of the Ottawa and Chippewa,* 58; Pokagon, *Ogimawkwe Mitigwaki,* 95; Ella Petoskey, "Chief Petoskey, " *Michigan History* 13 (July, 1929): 443–48; Verwyst, "Geographical Names," 396; Baraga, *Otchipwe Language* 2:353.
10. Romig, *Michigan Place Names,* 451; biographical details from Peckham, *Pontiac,* passim, and Parkman, *Conspiracy of Pontiac,* passim; Thwaites, *Early Western Travels* 1:170; Robert Rogers, *A Concise Account of North America* (reprint, New York: Johnson Reprint Corp., 1967), 240 *ff.*; name interpretations from Peckham, *Pontiac,* 19, Dwight H. Kelton, *Indian Names of Places Near the Great Lakes* (Detroit: Detroit Free Press Printing Co., 1888), 46, Gagnieur, "Indian Place Names," (1919), 416.
11. Weissert, "Indians of Barry County," 324–25; Baraga, *Otchipwe Language* 2:366.
12. Leo C. Lillie, *Historic Grand Haven and Ottawa County* (Grand Rapids: A. P. Johnson, 1931), 166; Romig, *Michigan Place Names,* 576; Baraga, *Otchipwe Language* 2:391.

13. Tanner, *Narrative of Captivity,* 19, 130; Blackbird, *History of the Ottawa and Chippewa,* 27; Albert Baxter, *History of the City of Grand Rapids, Michigan* (New York & Grand Rapids: Munsell & Co., 1891), 29; Everett, *Memorials,* 282; Romig, *Michigan Place Names,* 586; Etta Smith Wilson, "Life and Work of the Late George N. Smith, Pioneer Missionary," *MPHSC* 30 (1905): 190–212.

# V

# Potawatomi Personal Names

According to the memoir of Richard Fowler, who came to Hillsdale County from Massachusetts in 1834,

> . . . while my brothers were skating on a small pond near our place an Indian came to them and told them that if they would go with him he would show them a big Beese (which was the Indian name for lake); they did so, and since then the lake has been known as Baw Beese.

Fowler was mistaken in asserting that *beese* meant lake. A small inland lake was called *sâgaigan,* in the dialects or languages of the tribes of the Three Fires. The name Baw Beese is most probably a corruption of Ojibwa *babâwasse,* "a clear piece of land, a clearing or a lake, is seen through the woods." The local Potawatomi chief in Hillsdale County was called Baw Beese by the white settlers. According to the account just quoted, Baw Beese's band numbered about 150, and they often camped near Bird Lake.

Known as a peace chief, Baw Beese "was a tall, handsome, well-proportioned man. In business transactions his word could always be relied on," according to Archie Turrell, an early settler. The chief's name remains attached to the lake and to a village situated upon its shore.[1]

Bertrand is not an Indian name, but it was the European name of a notable mixed-blood family in Berrien County. Joseph Bertrand was a French-Canadian who married Mona, later called Madeline, the daughter of the Potawatomi chief Topenebee. About 1808 he established a trading post on the St. Joseph River at the point where it was crossed by the Great Sauk Trail, later called the Chicago road. This place was long called *Parc aux vaches* (cow pasture).

The Bertrands became parents of ten children who, with their mother, were to prosper from treaty benefits granted them in return for their father's exertion of influence with the Indians. In the Treaty of Chicago, August 29, 1821, Madeline Bertrand was given a section of land at the site of her husband's trading post, while five of her children received half-sections at the portage of the Kankakee, near South Bend, Indiana. In a treaty signed at the Mississinewa River of Indiana on October 16, 1826, two more Bertrand children were granted one section each, to be located at the direction of the president. In a treaty signed at the mission on St. Jospeh River, September 20, 1828, Madeline was given another section of land on the St. Joseph. In the Chicago treaty of September 27, 1833, Madeline and seven of her children each received $200, while three others received $100 each.

Most of the Bertrand children went to Kansas with the Potawatomi in the late 1830s. The signature of Joseph Bertrand, Jr., appears as a witness to a treaty signed at Potawatomi Creek, Kansas, June 17, 1846. B. H. Bertrand signed a treaty at Kansas River, November 13, 1861.

Madeline died in Michigan in 1847 and was buried at the present town of Bertrand, Berrien County. Joseph Bertrand, Sr., went to the Kansas reservation about 1858. He died in 1862 and was buried at St. Mary's, Kansas.

The town of Bertrand in Niles Township, and the adjacent township of Bertrand (the two once composed a single township) in Berrien County, are reportedly named for the senior Bertrand, although the names might broadly be considered as a memorial to the entire family.[2]

Cocoosh Prairie in Branch County allegedly preserves the name of a Potawatomi Indian whose name appears as Kokoosh in the Treaty of Saginaw, September 24, 1819. The name is Ojibwa and Potawatomi for "Hog." According to the account of Franklin Everett, Cocoosh was a black who, as a boy of twelve years, was captured by Indians during the War of 1812, and upon reaching maturity was made a chief. According to Everett, Cocoosh was the Indian name for Lyons, in Ionia County. Near Lyons is the junction of Maple River with the Grand. Along the Maple River, at a place called German Flats, was a Potawatomi village headed, in the early 1830s, by one Mocottiquahquash, or "Old Hog," according to Everett. The first half of this name, *mocotti,* does not mean "old," but seems to be one spelling of the Potawatomi word for "black." The remaining part, *quahquash,* is identical with Cocoosh,

Kokoosh, and so on, signifying "hog." The name thus translates as "Black Hog," giving some foundation to the story that this man was black.

There is some obscurity here, since Ionia County is a long way from Branch County, where the name Cocoosh is preserved. Moreover, the treaty of 1819 was with the Ojibwa, not the Potawatomi. In addition, one Coe-coosh had his portrait painted twice in 1845 by artist Paul M. Kane, who identified him as a resident of northern Wisconsin. He could be a different person, or he could have moved from one place and tribe to another, a common happening. However, the prairie could have been named for a different person or for the animal itself. The word was of course invented by the Indians only after they had seen hogs introduced by the French.

Cocoosh Prairie in Branch County is traversed by Hog Creek, undoubtedly named in translation from the Indian name.[3]

Mecosta County, containing a lake, township, and village of the same name, commemorates a local Potawatomi chief whose name is spelled this way in a treaty signed in Indiana on April 22, 1836. Jenks translates the name as "bear cub." If that is correct, the name has been altered from *maconse, makons, maw-kwans, maw-quans,* and so on, representing "little bear" or "young bear" in Ojibwa, Ottawa, and Potawatomi. The ending *ns, s,* or *se* added to a noun signify the diminutive. The name decidedly does not mean "Big Bear," as one source gives it. Conceivably it could be a corruption from *makoshtigwan,* "bear's head."

Moc-conse is listed as a signer of a treaty with the Potawatomi at St. Joseph on September 29, 1828. This could be the same individual whose name appears as Mecosta in the treaty of 1836. However, Schoolcraft also mentions "Maconse (the Little Bear)" as a chief of the Swan Creek band of Chippewa (Saginaw County) in 1838. Some treaty clerk gave a Scottish twist to the name by turning it into "McCoonse" in a treaty signed at the Sac and Fox agency, in Kansas, on July 16, 1859. It appears that neither the Potawatomi or Chippewa bearers of the name ever resided in Mecosta County. From the spelling, it is likely that the name honors the Potawatomi chief Mecosta.[4]

Monguagon Township in Wayne County once contained a village called Monguago, later renamed Trenton. It was the site of a Potawatomi attack on American troops on August 9, 1812. Monguago was a Potawatomi chief whose name is found in two treaties. The first, signed at Chicago on August 29, 1821, awarded "To Monguago, one-half section of land, at Mish-she-wa-ko-kink." His name appears again as Mo-gua-go among the signers of a supplementary article signed the day after the Chicago treaty of September 26, 1833, although he did not sign the treaty itself.

Monguago removed to Nottowassepee reservation in St. Joseph County in 1821 or later and lived between Leonidas and Mendon. In 1833 he was succeeded as a band chief by his son John. He and the other Indians resisted the removal called for by the Chicago treaty and were removed to Kansas by military force in 1840.

Another chief, called "the King of Monguagon," probably an ancestor of the one described here, was prominent during the French occupation of Michigan. A document signed by Governor Vaudreuil at Montreal, September 4, 1744, reads:

> We on good evidence, which has been produced to us, of the religion, the zealous attachment to the French, and the devotion to the service of the King of Monguagon, of the village of the Pottawatamies, have nominated and appointed him chief of the said Pottawatamies, with authority and command over the warriors of said village.

The terminal *gon* or *on* in the township name is a locative suffix. The name Monguagon may be derived from the Ojibwa word *mang,* "loon," or *mangon,* "large canoe."

Henry R. Schoolcraft was deceived about this name by Henry Conner, a subagent to the Saginaw Chippewa. Conner, who appears to have been a windbag and fabricator, told Schoolcraft that the place was named for Mo-gwaw-go, signifying "dirty backsides," which was allegedly the name of a Wyandot who died at Monguagon. There is not a shred of evidence in support of his tale.[5]

Leopold Pokāgón (1775?–1841) was during his adult lifetime second in importance only to Topenebee among Michigan Potawatomi chiefs. Reportedly born of a Chippewa father and an Ottawa mother, he became a Potawatomi by marriage to a Potawatomi woman. Owing to early contacts of his people with French priests, he was a devout Catholic and visited Detroit to ask for a resident priest for his village near Bertrand in Berrien County.

In response, Father Stephen Badin was sent to them, and a chapel was built.

Pokagon signed three treaties. In the first, at St. Joseph, Michigan, September 20, 1828, his name is spelled Po-ka-gon. In a treaty signed at Tippecanoe, Indiana, October 27, 1832, he and his wife were granted two sections of land, including his homesite. His name in that document is written Po-ka-gou. The next year, at the Chicago treaty of September 26, 1833, wherein the Potawatomi sacrificed their last lands in Illinois and Wisconsin, Pokagon had to give up the lands he was granted the year before; in lieu of them, he was to receive $2,000. Po-ka-gon is listed as the signer of amendments added to that treaty on September 27, although his name does not appear as a signer of the main document. Using the cash granted to him, he bought 712 acres in Cass County, Michigan, and lived there until his death in 1841.

At that time his youngest son, Simon, was eleven years of age. At age fourteen he attended school at Notre Dame, Indiana, remaining three years, after which he studied at Oberlin College in Ohio for one year. Next he spent two years in the academy at Twinsburg, Ohio, where he met Andrew J. Blackbird and Francis Petoskey, well-known Ottawa. In later years he made two trips to Washington to confer with President Lincoln concerning payments due to the Potawatomi.

As educated Indians were uncommon in his time, Simon became a lecturer and writer of note, publishing articles in *Harper's* and the *Review of Reviews.* He was a speaker at the Chicago World's Fair of 1893. His autobiographical novel, *Ogimawkwe Mitigwaki (Queen of the Woods)* was published shortly after his death, which took place in Allegan County on January 28, 1899.

The several places called Pokagon are named for Leopold Pokagon. In Michigan, these include the village and township of Pokagon, Pokagon Prairie, and Pokagon Creek, in Cass County. Outside Michigan are Pokagon State Park in Steuben County, Indiana, and the village of Pokagon in Kingfisher County, Oklahoma.

The name Pokagon appears to be a simplification of the Potawatomi term *opukeginima,* meaning "rib," although Simon Pokagon reportedly said it meant "protector."[6]

Shabbona, in Sanilac County, bears the name of a Potawatomi band chief (1775–1859). Different authors have claimed that his birthplace was in Quebec, Ohio, and Illinois. However, he lived in Michigan some time prior to the War of 1812, after which he moved to Illinois. Shabbona was born of an Ottawa father and a Seneca mother and was said to be a nephew of Tecumseh and a grandnephew of Pontiac. He married a Potawatomi woman and joined her tribe after the War of 1812.

Shabbona fought in the battle of Tippecanoe, Indiana, on November 7, 1811, and later for the British in the war that began the next year. After Tecumseh was killed in Ontario at the battle of the Thames on October 5, 1813, Shabbona transferred his allegiance to the United States. For most of his remaining years, he lived in Illinois. He persuaded the Potawatomi to remain neutral in the Winnebago troubles of 1827, and again during the Black Hawk War of 1832, when he also warned settlers of impending Indian attacks.

Two sections of land were reserved for Shabbona at Shabbona Grove in DeKalb County, Illinois, by the Treaty of Prairie du Chien in 1829. He migrated to Iowa in 1836, following the removal of his tribe to that state, but later returned to his grove. During a second visit with his tribe, after their removal to Kansas in 1848, his land was declared vacant and sold to speculators. Upon his return, he was beaten and turned away by a new occupant of the land. White friends provided him with forty acres near Morris, Illinois, where he lived on a small pension until his death on July 27, 1859.

In five treaties and related documents that contain Shabbona's name, it is spelled five ways. Its origin and meaning are not satisfactorily explained. Juliette Kinzie said the name came from that of one of his alleged birthplaces, Chambly, near Montreal. She called the name a corruption of the French *champ-de-blé*, "field of wheat." Another view maintains that Shabbona was named for a Captain Jacques de Chambly. Although Shabbona's grandniece, Frances Howe, denied that his name was of French origin, such origin is suggested by the appearance of his name as Chamblee in the St. Louis treaty of August 24, 1816. Still another view is that the name is of Potawatomi derivation, signifying "built like a bear," although it does not contain the Potawatomi word for bear, *makwa*.

Other places named for Shabbona are in Illinois and Iowa.[7]

During the early nineteenth century a local Potawatomi chief in Cass County followed the ancient custom of keeping the hair shaved from most of his scalp, and the whites called him Shavehead. Reports of him from old settlers seem to reflect a certain hostility that may or may not be justified. He is portrayed as a terror in his vicinity, feared by both whites and Indians. It is said that he hated whites, participated in many battles, and boasted of the scalps he had taken and wore as trophies.

The report that he owned a string of ninety-nine white men's tongues was termed a "great exaggeration" by Frances Nichols; actually, it is probably complete fiction born of the Indian-hating passions of frontier days. It is further reported, more credibly, that Shavehead established a ferry at the St. Joseph River in Cass County and "demanded tribute" from settlers wishing to get to the nearest grist mill by that route. There are no reports of free ferries being operated by anyone in those days. It is further reported that a settler caught him unawares and inflicted a severe beating, which "had the effect of curing his depredations but making him more sullen."

There are three differing reports of the manner of Shavehead's death, two of which maintain that he was killed by white men. The third story says he died of old age and poverty, that his grave was opened by whites, who cut off his head, and that in 1889 his skull was in the collection of the pioneers of Van Buren County.

A post office in Cass County was called Shavehead from 1858 to 1888. The name continues on a lake and prairie in Cass County.[8]

A monument at Tekonsha in Calhoun County commemorates Tekonquasha, a local Potawatomi chief who lived from 1768 to 1825. Both the village and township of Tekonsha are named for him. The full analysis of this name by Father Verwyst reads:

> Corruption of *attikonsan* (little caribou, or rein-deer), or *at-tik-on-saun.* Attik is rein-deer; attikon is the diminutive form; *attikon-san* is plural.

The name is clearly akin to that of Crispus Attucks, of Boston massacre fame, who was said to be part Indian and part black and whose name meant "little deer."

Tekonsha is an obscure figure; his name is not found in any treaties. His name on the map of Calhoun County dates from 1836, when it was adopted by the post office.[9]

In a treaty signed at Detroit on November 17, 1807, with territorial governor William Hull, four Indian tribes gave up their claim to northwestern Ohio and southeastern Michigan, but several small tracts were reserved for their use. One of these reservations consisted of "two sections of one square mile each, at *Tonquish's* village, near the river *Rouge.*" In the list of signers, the name of the chief of this village is spelled Tonguish, and he is listed as a Chippewa, as he was in an earlier treaty signed July 4, 1805. In the 1807 treaty, the apparently confused clerk listed a Toquish among the Potawatomi signers, but it is an obvious duplication. In treaties signed at Greenville, Ohio, July 22, 1814, and September 29, 1817, at the site of Toledo, the chief's name among the Potawatomi signers is spelled Tonguish.

In 1827, twenty years after the creation of the reservations in southeastern Michigan, all, including Tonquish's, were ceded to the government, along with most of southern Michigan, in return for a consolidated reservation on the Nottowassepee. The two sections of land "at Tonguish's village, near the river Rouge" were listed among the surrendered tracts, but the chief did not sign the document.

An early settler reported in 1885 that both Tonquish and his son had been killed in a skirmish with white settlers in 1819, but no verification of this story has been found. The name of this dispossessed chief remains on Tonquish Creek in Washtenaw and Wayne counties, which joins the River Rouge at Westland. Its meaning has not been determined.[10]

During the first third or more of the nineteenth century, no Potawatomi chief had more authority and distinction than Topenebee or Topinabee. His home band lived on the St. Joseph River in Berrien and Cass counties, but the town of Topinabee, which commemorates him, is in Cheboygan County, far to the north. It was named in 1881 by a hotel operator who platted the village. In Pokagon township of Cass County, Topinabee's Spring is his only other place-name memorial, aside from some street names in Niles and elsewhere.

Topenebee (as his name is spelled in the Chicago treaty of 1833 and in Hodge's *Handbook of American Indians*) must have been born about 1740, if we may credit missionary Isaac McCoy's estimate that the chief was over eighty years old in 1821. The date of his death is in dispute; some, including McCoy, place it in 1826, others in 1840. It is certain that he had a son of the same name, which is listed in the Treaty of Pottawatomie Creek, Kansas, signed June 17, 1846.

The elder Topenebee signed the Treaty of Greenville with Anthony Wayne on August 3, 1795. His name appears there as Thupenebu and is spelled five more ways in the eleven treaties that bear this name from that time until 1833. Topenebee responded favorably to the efforts of Tecumseh to unite the Indians against land cessions in the first decade of the nineteenth century. He was with the hostiles in the attack on the retreating garrison of Chicago's Fort Dearborn on August 15, 1812, but reputedly exerted himself to preserve the lives of the survivors.

After the war Topenebee followed a peaceful path and yielded to the pressure to sign treaties ceding huge chunks of land. In return, the treaty signed at St. Joseph on September 20, 1828, gave him (or his son?) a lifetime annual pension of a hundred dollars worth of goods. A section of land was reserved for "To-pen-i-be-the, principal chief of the said tribe, during his natural life," in the Treaty of Tippecanoe, October 27, 1832. It is reported that he continued to occupy his village on St. Joseph River, opposite Niles, Berrien County, until 1833. It is further said that he never went to the western reservations, but spent his last years in Silver Creek Township, Cass County.

According to Jacob Piatt Dunn, Topenebee's name signified "Quiet Sitting Bear." Elsewhere it is translated as "Peacemaker."[11]

The names of Wabasis Lake, Little Wabasis Lake and Wabasis Park, as well as Wabasis Creek, which joins Flat River in Montcalm County, are said to commemorate a Potawatomi chief who lived in that area. Sources translate his name as "White Swan," although "Little Swan" would be more accurate (*wa-bi-si,* "swan" plus terminal *s,* a diminutive).

It is further remarked that Wabasis was the adopted son of Wabwindego. Virtually everything else said about him is unverified legend. It is claimed that he signed away tribal lands in a treaty of 1826, by one account, or 1833, by other accounts. His name cannot be found in either of those treaties, or in any other, unless it is mangled beyond recognition. The name Wah-bou-seh in the Chicago treaty of September 26, 1833, is probably a variant rendering of the name of Wabonsee, or Waubaunsee, who lived in Illinois.

It is further alleged that Wabasis received a pot of gold for his acquiescence in land cessions and did not share it with his tribe, but buried it. For this he was allegedly ostracized and compelled to live within the limits of a garden tract on Wabasis Lake. Eventually, the story goes, he was lured to a corn feast at Plainfield, where he was made drunk and then murdered. He was buried with his head above the ground, and tobacco and food were left at the gravesite for many weeks. Finally, his skull ended up in an unnamed museum in New England. The story has the earmarks of white manufacture.[12]

Weesaw Township in Berrien County is named for a Potawatomi subchief who married a daughter of Topenebee. His village was on St. Joseph River two miles downstream from Niles, Berrien County. Anecdotes concerning him report that he was fond of adornment and friendly to whites. On one occasion in 1827 he provided food for the surveying party of Orlean Putnam when a supply party failed to find them. In 1832 he moved to Cass County and was reportedly shot "soon after" by his drunken son.

Weesaw's name appears in five treaties between 1821 and 1837, in which the spelling variants include Weesaw, Wesaw, We-is-saw, We-wis-sah, and Wisaw. In one of them Louison is given as his alternate name. In the Treaty of Tippecanoe, October 27, 1832, a half-section of land was reserved to him. In a treaty signed at Chippewaynaung, September 23, 1836, four sections of land were reserved for "the band of We-saw."

There is a Wesaw Creek in Miami County, Indiana, which was supposedly named for a Miami chief, but since no such name can be found in any Miami treaty, it was probably named for the Potawatomi chief, whose land claims extended into Indiana. Jacob P. Dunn called *Wesaw* the Miami name for the gallbladder of an animal, while J. N. B. Hewitt called it a Potawatomi name meaning "he, the torchbearer." The latter interpretation corresponds more closely to vocabulary sources.[13]

The town and township of White Pigeon in St. Joseph County honor the memory of a Potawatomi Indian about whom contradictory stories are told. He has been called a prophet for the Shawnee statesman Tecumseh. In 1812 he was described by Thomas Forsyth, Indian agent at Peoria, Illinois, as friendly to the Americans, yet in that same year he was the courier who brought a war message from the Shawnee to the Miami. It is further said that he warned settlers of a threatened Indian attack in 1812. A wilder story holds that about 1830, White Pigeon gave his life to save the settlement at the site of the present town of White Pigeon, although there were no Indian hostilities at that time. It is said that he died in 1830 at about thirty years of age, yet he is also reported to have aided whites two years later during the Black Hawk War, which never reached Michigan. Moreover, the name "Wapmeme, or White Pigeon," is signed to the first Treaty of Greenville, August 3, 1795, before the alleged time of his birth. That name appears again, in several spellings, in treaties signed in 1818, 1833, and 1846. The last document may have been signed by a son and namesake. In 1909 a monument to White Pigeon was erected on the edge of the town that has borne his name since 1833.

In Steuben County, Indiana, Pigeon Lake and the Pigeon River are also named for White Pigeon. The river flows past the town of White Pigeon, Michigan, and joins the St. Joseph River, a tributary of the Elkhart. There is also a town called White Pigeon in Whiteside County, Illinois, apparently named by settlers from Michigan.

The Indian name of White Pigeon, *Wapmeme,* and its variants, has also been given to the dove (*wabo-meme,* pronounced *wah-bo-mee-mee*).[14]

## Notes

1. Archie M. Turrell, "Some Place Names of Hillsdale County," *Michigan History* 6 (1922): 574; V. L. Moore, "Baw Beese Lake," *Michigan History* 16 (Spring, 1932): 334–47; Baraga, *Otchipwe Language* 1:153, 2:61–62.
2. Winger, *Potawatomi Indians,* 115–19; Fox, "Place Names of Berrien County," 32–33; Kappler, *Indian Treaties,* 200, 277, 295, 412, 560, 828.
3. Everett, *Memorials,* 73, 279; Kappler, *Indian Treaties,* 188; Baraga, *Otchipwe Language* 1:134; Rev. Maurice Gailland, "English-Potawatomi Dictionary" (Bureau of American Ethnology catalog no. 1761, ca. 1870, MS, microfilm), 148; Edmunds, *The Potawatomis,* 61; James A. Clifton, *The Prairie People* (Lawrence: Regents Press of Kansas, 1971), 111, 310; Romig, *Michigan Place Names,* 268; E. Wallace McMullen, "Prairie Generics in Michigan," *Names* 7 (September, 1959): 190.
4. Kappler, *Indian Treaties,* 296, 459, 795; Jenks, "County Names of Michigan" (1912), 467; Blackbird, *History of the Ottawa and Chippewa,* 107, 121; Pokagon, *Ogimawkwe Mitigwaki,* 108, 140; Baraga, *Otchipwe Language* 1:25; Romig, *Michigan Place Names,* 360; Schoolcraft, *Personal Memoirs,* 585.
5. Kappler, *Indian Treaties,* 199, 411; Charles Weissert, "Indians in 'Bitter Minority' in District until Expulsion," *Kalamazoo Gazette,* December 29, 1946; *MPHSC* 8 (1885): 459; Edmunds, *The Potawatomis,* 188; Baraga, *Otchipwe Language* 2:217–18; Schoolcraft, *Personal Memoirs,* 576.
6. For treaties signed by Leopold Pokagon see Kappler, *Indian Treaties,* 296, 374,

411–12; the name interpretation is from Gailland, "English-Potawatomi," 299; biographical data are from Cecilia B. Buechner, *The Pokagons* (Indianapolis: Indiana Historical Society, 1933); Frederick J. Dockstader, *Great North American Indians* (New York: Van Nostrand Reinhold, 1977), 215–17; George R. Fox, "Place Names of Cass County," *Michigan History* 27 (Summer, 1943): 477; Greenman, "Indian Chiefs," 229–30; Hodge, *Handbook of American Indians* 2:274–75; Winger, *Potawatomi Indians,* 141–58.

7. Juliette A. Kinzie, "Chicago Indian Chiefs," *Bulletin of Chicago Historical Society* 1 (August, 1935): 109–10; Kappler, *Indian Treaties,* 133, 298, 404, 410, 560; Vogel, *Indian Place Names,* 132–35.
8. F. S. Nichols in Hodge, *Handbook of American Indians* 2:530; A. B. Copley, "The Pottawattomies," *MPHSC* 14 (1889): 265–67; A. D. P. Van Buren, "Deacon Isaac Mason's Early Recollections of Michigan," *MPHSC* 5 (1882): 401–2.
9. FWP, *Michigan,* 502; Romig, *Michigan Place Names,* 551; Verwyst, "Geographical Names," 397.
10. Kappler, *Indian Treaties,* 78, 94, 107, 151, 283; Hodge, *Handbook of American Indians* 2:778; Romig, *Michigan Place Names,* 557–58; Melvin D. Osband, "The Story of Tonguish," *MPHSC* 8 (1885): 161–64.
11. Romig, *Michigan Place Names,* 558; FWP, *Michigan,* 492; Fox, "Place Names of Cass County," 478; "Topinabee," *The Totem Pole* 28 (March 3, 1952): 1–3; Clifton, *Prairie People,* 186; Kappler, *Indian Treaties,* 44, 65, 107, 201, 275, 295–96, 373–74, 403, 410, 560; Catherine Stewart, *New Homes in the West* (Ann Arbor: University Microfilms, 1966), 26–29; Hodge, *Handbook of American Indians* 2:785; Winger, *Potawatomi Indians,* 95–98; Dunn, *True Indian Stories,* 255–56; Howard S. Rogers, *History of Cass County Michigan from 1825 to 1875* (Cassopolis: W. H. Mansfield, 1875), 59–60.
12. Pokagon, *Ogimawkwe Mitigwaki,* 72; Baraga, *Otchipwe Language* 2:393; Kappler, *Indian Treaties,* 404; Baxter, *History of Grand Rapids,* 30; J. D. Dillenback and Leavitt, *History and Directory of Kent County, Michigan* (Grand Rapids: J. D. Dillenback, 1870), 75–76; Robertson M. Augustine, "Indians, Sawmills, and Danes" (Greenville, Mich.: Flat River Historical Society, 1971, typescript), 11–14 (special thanks to John R. Halsey, Michigan Historical Commission, for these citations).
13. Alfred Mathews, *History of Cass County, Michigan* (Chicago: Waterman, Watkins & Co., 1882), 47–48; Copley, "The Pottawattomies," 265; Fox, "Place Names of Berrien County," 6–35; Kappler, *Indian Treaties,* 315.
14. Romig, *Michigan Place Names,* 600; FWP, *Michigan,* 398; Glenn Tucker, *Tecumseh: Vision of Glory* (Indianapolis: Bobbs Merrill, 1956), 360; Hodge, *Handbook of American Indians* 2:945–46; Greenman, "Indian Chiefs," 245–46; Clarence E. Carter, ed., *Illinois, 1809–14,* The Territorial Papers of the United States, vol. 16 (Washington, D.C.: U.S. Government Printing Office), 251–52; Kappler, *Indian Treaties,* 44, 100, 413, 559; Pokagon, *Ogimawkwe Mitigwaki,* 215; Ronald Baker and Marvin Carmony, *Indiana Place Names* (Bloomington: University of Indiana Press, 1975), 129; Vogel, *Indian Place Names,* 166–67.

# VI

# Other Personal Names

One of the most notable Indian personal names that is preserved on the Michigan map belonged to one who never lived within the state. That is the Shawnee warrior Tecumseh (1768–1813), whose name was given in 1824 to the seat of Lenawee County and its township, and later to Tecumseh Park in Lansing. That is appropriate, for Tecumseh traveled about Michigan in his crusade to unite Indians against land cessions to encroaching whites. The state of his birth, Ohio, gave his name to nothing more significant than a hill on Mad River.[1]

Tecumseh was born at the Shawnee village of Piqua near present Springfield, Ohio, in 1768. When he was six years old, his father was killed by whites in the battle of Point Pleasant, West Virginia. When he was about twenty, his elder brother who had reared him was killed by Tennessee frontiersmen. In the battle at Fallen Timbers, Ohio, 1794, another brother was killed.

Sometime after the Treaty of Greenville (1795), Tecumseh began his campaign to unite the Indians, not for aggression, as has long been maintained, but for retention of the best of Indian life ways, and the rejection of the white man's liquor and religion. Above all, he insisted that the land was the common property of all Indians, and none had a right to sell it. He stopped the practice of torture, promoted monogamy, and encouraged agriculture and stock raising. In 1808, he and his twin brother, Tenskwatawa, "the prophet," established themselves with followers from several tribes at the old Indian village site at the junction of Tippecanoe and Wabash rivers in Indiana.

Tecumseh was not actually a chief of the Shawnees, but was the chosen leader of his followers from several tribes. During the time of his greatest activity, the Shawnee chiefs Black Hoof, Blue Jacket, and Colonel Lewis signed treaties, but Tecumseh consistently refused to do so, for treaties meant land cessions. Tecumseh stood up to the incessant pressure for land cessions that came from territorial governor William Henry Harrison. Harrison therefore decided to disperse the Tippecanoe settlement and in the fall of 1811 marched upon it with a force of more than nine hundred militia. Tecumseh was absent on a visit with the Creeks in the South (to which tribe his mother had belonged), and Tenskwatawa was forced into a confrontation without the assistance of his brother. At first seeking a parley, to gain time, the prophet launched a predawn attack on the invaders, November 7, 1811, but the Indians were repulsed and scattered. Their village, livestock, and corn were all destroyed.

The effect of this unnecessary invasion by Harrison was to unite all tribes of the Northwest against the Americans when war with Great Britain began the following year. Tecumseh led his followers into the British army, and reportedly was given the pay and uniform of a brigadier general. On October 5, 1813, as a participant in General Henry Procter's army, he met his old nemesis, General Harrison, on the battlefield near Chatham, Ontario, where he was killed. The loss of the great Shawnee leader so demoralized the Indians that many of them took no further part in the war, and some joined the Americans.

In later years, two notable politicians sought to capitalize on their campaign against Tecumseh. In 1836, Colonel Richard M. Johnson of Kentucky claimed that he was Tecumseh's slayer and was elected vice-president. In 1840, General Harrison, dubbed "Tippecanoe" for his campaign of 1811, was elected as the first Whig president.[2]

Besides Michigan, Tecumseh's name is on towns in Kansas, Missouri, Nebraska, Oklahoma, and Ontario. Of the meaning of Tecumseh's name, A. S. Gatschet wrote:

> The name of chief Tecumseh (in Shawnee Tekamithi or Tkamthi) is derived from nila ni tkamthka, "I cross the path or way" (of an animate being). By this is meant that the name belongs to a totem of one of the round-footed animals, as that of the raccoon, jaguar, panther, or wildcat, and not to the hoofed ones, as the deer. Tecumseh and his brothers belonged to the . . . "miraculous panther" totem. . . . Tecumseh's name has been variously translated in former times as "panther lying in wait," "crouching lion," and "shooting star." All

> these only paraphrase the meaning, but do not accurately translate or interpret the name. . . .
>
> The quick motion of a meteor was evidently likened to that of a lion or wildcat springing upon its prey, and the yellow color of both may have made the comparison more effective.[3]

Except for some names from literature and legend, the name of only one Indian who had nothing at all to do with Michigan history is honored by its present place names, and that is Osceola of the Seminoles (ca. 1803–38). Moreover it appears five times: as the name of Osceola County, a village in Houghton County, townships in Houghton and Osceola counties, and Oceola Township (thus spelled) in Livingston County. The name of Osceola is, moreover, found in more place-names throughout the country than that of any other Indian. Much of this popularity is due to the literature that glorified his heroic resistance to deportation from Florida, particularly Captain Mayne Reid's novel, *Osceola the Seminole* (1858).

Born in the Creek country of Alabama, Osceola was not a chief but by his teen years was already a distinguished warrior. In 1832 he refused to sign the Treaty of Payne's Landing, by terms of which the Florida Indians were to be transferred to Oklahoma. According to a disputed account, when General Wiley Thompson called a parley to secure the endorsement of the holdouts, Osceola drove his knife through the treaty, vowing never to accept it. Osceola was then seized and placed in irons. The determined warrior pretended a change of heart in order to win release, but then renewed his resistance in the swamps, together with chiefs Micanopy and Jumper.

When Major Francis Dade marched with a hundred soldiers to capture him, all but three were killed in a battle near the site of present Bushnell, Florida, December 28, 1835. In another engagement soon thereafter on the Withlacoochee River, Osceola was wounded but managed to escape. He was captured only through trickery. General T. S. Jesup convened a pretended peace parley, to which Osceola and several followers came under a flag of truce. Upon their arrival, all were seized and sent first to St. Augustine, and then to a dungeon at Fort Moultrie, South Carolina. Osceola, suffering from malaria, followed by quinsy, died within three months, on January 30, 1838. Before his death, his portrait was painted by George Catlin. The Seminole resistance continued until 1842, when the government gave up the expensive war, permitting some three hundred Seminoles to remain in the Everglades, where their descendants are now increasing.

The name Osceola has been spelled in many ways, but there is little disagreement as to its origin and meaning. It is derived from the Creek word *assiyahola,* meaning "black drink singer." Black drink was a caffeine-rich ceremonial beverage, with emetic properties, made from the leaves of species of holly trees, *Ilex cassine* or *Ilex vomitoria,* also called by the aboriginal name *yaupon.* The name of Osceola is on the map of at least a dozen states.[4]

## Notes

1. Martin, "Ohio Place Names," 276.
2. Tucker, *Tecumseh,* passim; Hodge, *Handbook of American Indians* 2:714.
3. A. S. Gatschet, "Tecumseh's Name," *American Anthropologist* 8 (1895): 91–92.
4. Florida Historical Society, "The Complete Story of Osceola," *Florida Historical Quarterly* 33 (January–April, 1955): 161–305; Edwin C. McReynolds, *The Seminoles* (Norman: University of Oklahoma Press, 1957), passim; Hodge, *Handbook of American Indians* 2:159; William A. Read, *Louisiana Place Names of Indian Origin,* Bulletin 19 (Baton Rouge: Louisiana State University, 1927), 45.

# VII

# Names from Literature and Legend

The American Indian influence on American literature has resulted in dozens of Indian names and some pseudo-Indian names reaching the map through that indirect route. During the late eighteenth century and all of the nineteenth, poets, novelists, and playwrights won fame with works built on Indian themes. It was a curious time, for the century in which the native people made their last stand against white conquest was also the century in which they aroused the most romantic and sympathetic interest among white writers and their readers.

James Fenimore Cooper's novels *The Last of the Mohicans* and *The Deerslayer* and the series of *Leatherstocking Tales* have attracted the most interest in the category of Indian fiction; John Augustus Stone's *Metamora* among dramas and Henry W. Longfellow's *The Song of Hiawatha* among poems were most influential and best remembered, but dozens more writers pursued the Indian theme. Phillip Freneau, Captain Mayne Reid, and Helen Hunt Jackson are but a few of them. Some writers, such as Mary Eastman and Henry R. Schoolcraft, also published Indian legends (not always fully authentic) that generated interest in the side of the Indian that was hidden from view in stories of frontier violence.[1]

Although immensely popular in their day, nineteenth-century literary works dealing with Indian themes are not highly regarded today. That is of no consequence to the purpose of this chapter, which is to show how these works made certain Indian names and some pseudo-Indian names into household words and caused their adoption as both personal and geographic names. Both Eastman and Longfellow can be credited with the popularity of Winona as a name. Mrs. Jackson caused the name of Ramona, a California Indian heroine, to be adopted as a place and personal name. Cooper is apparently the source of the nickname Hawkeye for the state of Iowa and also caused the wide adoption of Mingo as a place-name. Captain Reid popularized the name of Osceola. Some writers invented artificial Indian names, such as Metamora. That name was adopted in several states and remained a puzzle until its literary origin was unveiled. There are dozens of such artificial names on the map, some devised by Schoolcraft, that will be separately considered.

Apparently the only name from Cooper's novels that has been adopted in Michigan is Lake Horicon, in Otsego County. A French map of 1671 showed that name or something similar as belonging to an Indian tribe west of Lake Champlain, New York. Generally it is supposed that the name was a French corruption of the name Mohican. By Cooper's own account, he took the liberty of putting Horicon into the mouth of his hero, Natty Bumpo, as the aboriginal name for Lake George. He preferred this course, "instead of going back to the house of Hanover for the appellation of our finest sheet of water." Thus, says William M. Beauchamp, Cooper did not invent the name but transferred it. Its attachment to Lake George was not accepted, but the name was adopted for a lake and village in Warren County, New York, as well as for a town, marsh, and wildlife refuge in Dodge County, Wisconsin, besides our lake in Michigan.[2]

No work of literature has equaled Longfellow's *Song of Hiawatha* (1854) in its influence on the map. The Cambridge poet sought authentic aboriginal names, mostly Ojibwa, not only for his human characters, but also for the flora, fauna, spirits, and natural phenomena of sky and water. Most of these he learned from the works of Henry R. Schoolcraft, the early American ethnologist whose influence on Michigan history and its place-names is especially notable. Longfellow, like Schoolcraft, was willing to mix language and cultural facts arbitrarily to achieve a desired result. Thus the name Hiawatha, of Iroquois origin, was chosen by him as a name for his hero, who was actually the Ojibwa deity Manabozho. Apparently the choice was made solely for euphony.

The authenticity of the historic Hiawatha is questioned by some but was accepted by Henry L. Morgan and Horatio Hale, both of them esteemed scholars of Iroquois life. By the best accounts, the real Hiawatha, who flourished about 1570, was a Mohawk sachem, spokesman, and aide to Dekanawida, a reformer and peacemaker who united five mutually antagonistic Indian nations into the

League of the Longhouse, the Iroquois. According to Morgan, Hiawatha, after his death, became a part of the Iroquois pantheon. Longfellow borrowed nothing but the name for his tale. All the feats of Longfellow's hero are those of Manabozho, a spiritual figure of the Ojibwa of Lake Superior, son of a father in the sky and a woman of the earth. His name, spelled Manabezho, is on a waterfall in Presque Isle River, Gogebic County.

In Michigan the name of Hiawatha is given to a national forest totaling 839,910 acres sprawled across several counties of the Upper Peninsula. Hiawatha is also the name of a township and a tiny village in Schoolcraft County. The village was founded in 1893 by socialists and populists as a cooperative colony, but it soon failed because of internal dissension. Small streams in Luce, Chippewa, Mackinac, and Schoolcraft counties also bear the name Hiawatha.[3]

There are three versions of the meaning of Hiawatha and also several spellings of the name. Morgan interpreted the name in Seneca dialect as "the man who combs," from the tradition that Hiawatha combed snakes from the head of an Onondaga hero named To-do-da-ho, who, like Medusa of antiquity, was covered with entwined serpents. J. N. B. Hewitt, the Seneca ethnologist, claimed that the name signified "he makes rivers." To Schoolcraft, the name denoted "a person of very great wisdom."[4]

Hiawatha's mother in the poem was Wenonah, an earthbound woman who was impregnated by a spirit, Mudjekeewis, the West Wind. Although she was clearly an Ojibwa and daughter of Nokomis (grandmother), Wenonah's name is the Dakota name for a first-born daughter.[5] As a name it first came to notice in William Keating's account of Long's expedition (1825), and again in Mary Eastman's *Dahcotah, or Life and Legends of the Sioux* (1849). Both narrate a legend of a Sioux girl who committed suicide by throwing herself from a bluff along the Mississippi to avoid marrying a man chosen for her by her parents. Of course, these happenings are entirely unrelated to the Wenonah in Longfellow's story, who died giving birth to Hiawatha.[6]

Each narrator of the Wenonah story spells the heroine's name in a slightly different way. In Eastman it is Wenona, in Keating Winona, and in Longfellow Wenonah. All of these spellings and others are found in the names of the two dozen places named for her in eighteen states.

Winona is the name of a small village in Houghton County, named in 1898 for a copper-mining company. Formerly there was a village called Wenona in Bay County, but it is now a part of Bay City. There are Winona lakes in Branch and Houghton counties.[7]

Hiawatha courted and won for his wife the maiden Minnehaha, daughter of a Dakota arrowmaker, who named her.

> And he named her from the river,
> From the water-fall he named her,
> Minnehaha, Laughing Water.

Although Longfellow translated her name as "Laughing Water," in truth it is a generic name, in Santee Dakota, for any waterfall. Its literal meaning is "curling water." Minnehaha Creek in Emmet County is named from the Longfellow poem, since all Dakota names in Michigan are introduced from outside the state.[8]

Hiawatha's grandmother, who reared him after the death of his mother, Winona, was called Nokomis. It is the Ojibwa word for grandmother. Undoubtedly the poem is responsible for the name of Nokomis Falls in Gogebic County and probably also for that of No-ko-mos Lake in Kent County.[9]

"Iagoo the great boaster," the storyteller at Hiawatha's wedding feast, is the source of the name of Iagoo Falls in Presque Isle River, Ontonagon County.

The name Keewayden, for the Northwest Wind in *Hiawatha*, has been given to a lake on the line of Baraga and Marquette counties. Variant spellings of the name—Kewadin in Antrim County and Keewahdin in St. Clair County—have other origins.[10]

A passage in *Hiawatha* reads:

> Sounds of music, words of wonder;
> "Minne-wawa!" said the pine trees.

The glossary of the poem lists *minne-wáwa* as "a pleasant sound, as of wind in the trees." This word is not in Baraga's dictionary; clearly it is the source of the name of Minnewawa Falls in the Presque Isle River, Ontonagon County.

When Hiawatha underwent a four-day fast prior to his wrestling match with Mondamin, the maize spirit, he spent the second day wandering "Through the Muskoday, the meadow," observing the wild food plants. That seems to be the likely source of the name of Lake Muskoday, Wayne County.

"In the Vale of Tawasentha," goes the Hiawatha story, "Dwelt the singer Nawadaha." It was he who "sang of Hiawatha" and told the stories here recorded by Longfellow. His name is preserved in that of Nawadaha Falls in the Presque Isle River, Gogebic County. Its meaning is not recorded, but it could be a corruption from Ojibwa *nawadjiwan,* "in the midst of a rapid."[11]

Onaway, the name of a town and state park in Presque Isle County, is "said to be the name of an Indian maiden," according to one writer. So often is this explanation given for Indian names that it should never be accepted without confirming evidence. The name Onaway appears on a town in Idaho, as Onawa on an Iowa town and a lake in Maine, as Onoway on a town in Alberta, Canada, and as Onway on a lake in New Hampshire. It is sometimes translated as "awake," "awaken," and "wide awake" (Stewart). Although these explanations conform to the *Hiawatha* glossary, they have no support in Ojibwa sources. In Baraga's dictionary, the term for "I awake" is *nin goshkos.* Onaway is a name from *Hiawatha,* however. At the hero's wedding feast, Chibiabos sang:

> Onaway! Awake, beloved!
> Thou the wild-flower of the forest!
> Thou the wild-bird of the prairie!

A misunderstanding led to the assumption that "Awake" was the English meaning of Onaway. However, it is probably the name of an imaginary person to whom the song of Chibiabos is addressed. Its meaning is undetermined.[12]

Clearly borrowed from the Hiawatha story is the name of Osseo, a village in Hillsdale County. It has been called the name of an "Indian chief,"[13] an explanation that appears with tiresome frequency in place-name writing. However, the name of no such chief can be located in the standard sources. Osseo is the son of the evening star, whose story is told in Hiawatha by Iagoo.

> And he said in haste: "Behold it!
> See the sacred Star of Evening!
> You shall hear a tale of wonder,
> Hear the story of Osseo!
> Son of the Evening Star, Osseo!"

The name has also been adopted for places in Minnestoa, Wisconsin, and Ontario. Such widely scattered appearances of a name frequently indicate a literary origin.

Also among the personal names from *Hiawatha* that appear on the Michigan map are those of the spirits of winter and spring.

> Mighty Peboan the Winter,
> Breathing on the lakes and rivers,
> Into stone had changed their waters.

The name is perpetuated on Peboan Creek in Chippewa County. In contrast to that grim reminder of northern winters, "Segwun, the youthful stranger," the spirit of spring, relates:

> And where'er my footsteps wander,
> All the meadows wave with blossoms,
> All the woodlands ring with music,
> All the trees are dark with foliage!

Spring's name is preserved in that of the unincorporated community of Segwun in Kent County on the south edge of Lowell.[14]

The name of Ponemah Lake in Genesee County is taken from the Ojibwa term used in *Hiawatha* to denote the hereafter, the home of departed souls. Ponemah is mentioned four times in the poem: when the spirit of Chibiabos is called from the grave, when the ghosts haunt Hiawatha during his journey homeward to Minnehaha's side, upon the death of Minnehaha, and at the final departure of Hiawatha.

> To the regions of the home-wind,
> Of the Northwest wind, Keewaydin,
> To the Islands of the Blessed,
> To the Kingdom of Ponemah,
> To the land of the Hereafter!

John Tanner translated the Ojibwa word *pon-ne-mah* as "hereafter," and the Ottawa *paw-ne-maw* as "by and by."[15] Ponemah is the name of an Ojibwa village on Red Lake Reservation, Beltrami County, Minnesota. There are also villages named Ponemah in Illinois, New Hampshire, and Manitoba.

Hiawatha sat one day listening to his father's boasts:

> Then he said, "O Mudjekeewis,
> Is there nothing that can harm you?
> Nothing that you are afraid of?"
> And the mighty Mudjekeewis,
> Grand and gracious in his boasting,
> Answered saying, "There is nothing,
> Nothing but the black rock yonder,
> Nothing but the fatal Wawbeek!"

Hiawatha then broke the large rock into fragments and threw them at Mudjekeewis, blaming him for the death of Wenonah, Hiawatha's mother.

There can be little doubt, despite the minor spelling difference, that Wabeek Lake in Oakland County draws its name from the Hiawatha story. The same can be said for other variants of this name on the map elsewhere: Waubeek, Iowa, the township of Wawbeek in Pepin County, Wisconsin, and Wawbeek, Alabama.

The glossary of *Hiawatha* defines *wau-beek* as "a rock." Baraga's dictionary has *âjibik* as the word for rock. Ojibwa words for iron, tin, and copper all contain the termination *bik* (pronounced *beek*). The type of mineral or its color may be indicated by a prefix—for example, Tanner has *muk-kud-dah-waw-beek,* "Black stone."[16]

Longfellow scoured Ojibwa vocabularies for the names of animals, birds, fish, and insects, incorporating many of them into the Hiawatha story. Several are on the Michigan map, but we cannot always be sure whether they were borrowed from the poem, although identical spelling is sometimes a clue. The name of *Kenosha,* the pike, appears as spelled in *Hiawatha* on a lake in Newaygo County and a park in Kent County. Baraga's spelling of this word is *Kinoje;* Tanner has *Ke-no-zha.* The name of Kenosha, Wisconsin, however, is a translation of the name of Pike River, on which it is located.[17]

*Nahma,* the sturgeon in Hiawatha, is a place-name in Delta County, but there it appears to be a translation from the name of Sturgeon River, on which it is situated.

The name of Kenabeek Creek, in Gogebic County, appears to

be borrowed from *Hiawatha,* as are several other place-names in that county. In *Hiawatha,* Longfellow speaks of

> The Kenabeek, the great serpents,
> Lying huge upon the water,
> Lying coiled across the passage. . . .

In Baraga's dictionary, the equivalent word is *gĭnebig,* "snake."[18]

Opechee Point on Lake Superior, Houghton County, has a name from the Ojibwa word for robin, as used in *Hiawatha.*

> Sang the Opechee, the robin,
> "Happy are you, Laughing Water,
> Having such a noble husband!"

Another passage in *Hiawatha* reads:

> Music as of birds afar off,
> Of the whippoorwill afar off,
> Of the lonely Wawonaissa
> Singing in the darksome forest.

Although the spelling is slightly different, the name of Wawonissa Creek, a tributary of Little Carp River in Ontonagon County, is probably taken from that of the whippoorwill in *Hiawatha.* It is spelled the same way in the vocabulary of John Tanner's *Narrative of Captivity.*[19]

Leelanau County has one of the most melodious names in Michigan. The county occupies the peninsula dividing Grand Traverse Bay from Lake Michigan, and within it are Leelanau Township, Lake Leelanau, Lower Leelanau Lake, and the village of Lake Leelanau.

The name Leelanau is that of the heroine of an Indian love story told by Henry R. Schoolcraft, called "Leelinau or the Lost Daughter," in his *Algic Researches* (1839). In the story, Leelinau (so spelled) was the daughter of a great hunter who lived near the Porcupine Mountains on the shore of Lake Superior. A timid and solitary girl, she spent much time in a pine forest called Manitowak, or Spirit Grove. Although she shunned company, her parents arranged that she should marry an older man, the son of a nearby chief. Her resistance to the match failed to dissuade them. The night before the day chosen for the marriage, she dressed in her best clothing and decorations and announced that she was about "to meet my little lover, the chieftain of the green plumes, who is waiting for me at the Spirit Grove." Supposing this behavior to be harmless fantasy, her parents did not interfere. The girl did not return, and was never found.

Near the Spirit Grove one evening a fishing party saw a girl standing on the shore. Paddling toward her, they recognized the missing daughter, but she fled from them. Before she disappeared, the fishermen saw "the great plumes of her lover waving over his forehead, as he glided lightly through the forest of young pine."

Schoolcraft, who probably invented the story, suggested the girl's name, in revised spelling, for the county. One explanation of the name, "delight of life," is purely fanciful. Moreover, there is no *l* sound in Ojibwa. The closest approach to the name Leelanau in an Algonquian language appears to be the Montagnais word *laleu*, "seashore." Mentor Williams suspected that this story was chiefly Schoolcraft's invention, adding that "The geography . . . links it to contemporary Indian tales in popular magazines rather than to an authentic Indian source."[20]

Ontwa Township in Cass County was named, according to one account, for an Indian girl employed by Thomas H. Edwards, the first township clerk. If there was in fact such a person, her name is apparently much corrupted from its original form, and cannot be interpreted.[21] More likely the name arose, directly or indirectly, from a poem by Harry Whiting, *Ontwa, the Son of the Forest* (New York: 1823). In that case, it is not a female name, actual or intended. Since this poet is also the one who borrowed from Schoolcraft or Cass the name of Sannilac in another poem, this name, too, may have been suggested by Schoolcraft, who was acquainted with Whiting and wrote notes for the Sannilac poem.

Along Lake Michigan in Leelanau County is a range of sand dunes, of which the most prominent is called Sleeping Bear. From it is named Sleeping Bear Bay and Point on the lower Leelanau peninsula. As early as 1688 this place is shown as *L'Ours qui Dort* (the Sleeping Bear) on Franquelin's map. Antoine Raudot wrote in 1710: "There is a mountain there which the savages call the sleeping bear, because it is shaped like one. They [the Indians] say that after the flood the canoe which saved their fathers ran aground there and stopped." Still another account of this place was given by the missionary Pierre Charlevoix, on August 1, 1721.

> I perceived on a sandy eminence a kind of grove or thicket, which when you are abreast of it, has the figure of an animal lying down: the French call this the Sleeping, and the Indians the Crouching Bear.

Before the white men came, according to an Ottawa-Chippewa legend, a black bear with two cubs tried to swim across Lake Michigan, from the west side. When approaching the Michigan shore, the cubs tired and lagged behind. The mother bear climbed the dunes to await her cubs, but they never arrived. She remained on the dune as a patch of dark vegetation contrasting with the light colored sand. The cubs were transformed into the islands now called North and South Manitou (Spirit). On North Manitou today are the village of North Manitou and Lake Manitou. These islands, together with Fox and Beaver islands, once formed Manitou County, which was later divided between Leelanau and Charlevoix counties. The mainland dune area has recently become Sleeping Bear Dunes National Park.[22]

> Had Sannilac to gain a name?
> Was he unknown to warlike fame?
> The Wyandot tribe will answer, no!
> Ere yet this arm had strength to fight,
> This heart had all the warrior's might,
> And long'd to strike a Mingo low,
> And when, at last, by Tarhee's side
> I revel'd in the battle's pride,
> What warrior dealt a surer blow?

So goes a verse in Henry Whiting's poem *Sannilac,* published at Boston in 1831. It contains an appendix of notes by Lewis Cass, governor of Michigan Territory, and Henry R. Schoolcraft. The story is one of Wyandot retribution against the Mingo (Iroquois) in the days before white settlement. The Wyandots were led by Sannilac, a spirit warrior who won the love of a maiden named Wona by defeating the enemy. Whiting clearly explained in his preface that the purpose of the story was "not so much to fill up the outline of history, as to exhibit manners and customs, which are generally characteristic of the sons of the forest." Nothing is said about the origin and meaning of the name Sannilac, but it is obviously an artificial name, having no roots in either Algonquian or Iroquoian languages. It could have been suggested by Schoolcraft, whose penchant for manufacturing names is dealt with elsewhere. However, William Jenks held that Whiting obtained the name from Cass's manuscripts.

The county name, spelled Sanilac, was first used in a proclamation by Governor Cass on September 10, 1822, before the publication of Whiting's poem. The legislative act organizing the county was passed April 3, 1848, long after statehood was achieved. The village of Port Sanilac was named for the county in 1857.[23]

In New York, in mid-December, 1829, a play was performed that, according to Albert Keiser, "gave the Indian drama its greatest impetus and prolonged its popularity for two decades." Entitled *Metamora, or the Last of the Wampanoags,* it was written by twenty-five-year-old John Augustus Stone and starred popular actor Edwin Forrest. The story was patterned about the life of Metacomet, or King Philip of the Wampanoags, the second son of chief Massasoit, or Osamequin, who befriended the Pilgrims in 1621. Metacomet became chief of his tribe in 1672 and four years later led the unsuccessful rebellion known as King Philip's War, during which he was killed in the "swamp fight," August 12, 1676.

The popularity of this play caused the adoption of the name Metamora by a village and township in Lapeer County, Michigan, and later it was given to the Metamora-Hadley State Recreation Area a mile west of Metamora. The name was also adopted for places in Ohio, Indiana, and Illinois. It is possible that the name of Oceana County, Michigan, also came from the play, wherein Oceana is portrayed as a white maiden who was once rescued from a panther by Metamora, who repaid him by treating his wounded arm.

Although Metamora is a contrived name, its first half comes from the name of Metacomet and may signify "heart" (*metah*).[24]

In 1832 a British major, John Richardson (1796–1852) published a novel of Indian warfare entitled *Wacousta, or The Prophesy.* It went through numerous editions and was issued in Canada in

1868. In the novel, Wacousta is the Indian name of Reginald Morton, a fictional white "renegade" who was adviser to Pontiac in the rebellion of 1763. As his role is explained by literary critic Louise Barnett,

> this renegade, rather than the Indian leader Pontiac, is the intransigent spirit behind the frontier war of 1763. Endowed with the typical attributes of the stereotype, Wacousta out-Indians the Indians: he performs marvelous physical and military feats, devotes himself wholly to revenge, and is unsurpassingly bloodthirsty and unmerciful to captives.[25]

From this unsavory individual who never existed sprang several place-names. In Michigan, the name Wacousta was given in 1839 to a post office in Clinton County. Wacousta continues as a village, although its post office is long gone. As we have indicated before, names popularized in novels are often adopted in several places, and so Waucousta is also the name of a village in Fond du Lac County, Wisconsin, and of a township in Humboldt County, Iowa.

Helen Hunt Jackson (1832–85), an ardent defender of Indian rights, published her best known novel, *Ramona,* in 1884. It was a story of the mistreatment of California Mission Indians, and its heroine, Ramona, was modeled after a living Indian woman. Although her name was Spanish, the real and fictional Ramona was a Cahuila Indian. She was Ramona Lubo, who died July 21, 1922, and was buried in the Cahuila Indian cemetery in California's San Jacinto Mountains. Already notable for her earlier nonfiction book, *A Century of Dishonor* (1881), Mrs. Jackson drew national attention to the dispossession of Indians and won the denunciation of Theodore Roosevelt for her passionate defense of the native Americans. Owing to the popularity of her novel, the name Ramona was given to Ramona Park, a village in Emmet County, and to the village of Ramona in Newaygo County, which was named in 1904 by a resort owner and summer resident. Ramona is also the name of places in seven other states.[26]

Along Lake Michigan about seventeen miles north of downtown Chicago is the affluent suburb of Winnetka. When platted in 1857 it was called Wynetka, but when incorporated twelve years later the spelling was changed to Winnetka. The name was reportedly taken by Mrs. Charles E. Peck, wife of the village founder, from a novel she had read. She claimed it was an Indian word meaning "beautiful land" or "beautiful place." Neither the language of the name nor the name of the book has been identified, but it appears that the same mysterious novel is the source of the name of Winetka Point in Benzie County, Michigan. It appears to be a manufactured name based on aboriginal roots that were trimmed to fit someone's concept of euphony. The meaning, if any, may not be what it is believed to be. If the name is of Ojibwa origin, we must contend with the fact that *win* or *wini,* in compositions, means "unclean, impure." Winad-aki would mean "unclean ground."[27]

"There is no doubt," observed a western writer over a century ago, that "the Indians would be much amused if they could know what a piece of work we have made of some of their names."[28]

## Notes

1. Constance Rourke, *The Roots of American Culture* (New York: Harcourt Brace, 1942), 69–74; Benjamin Bissell, *The American Indian in English Literature of the Eighteenth Century* (reprint, Hamden, Conn.: Archon Books, 1968), passim; Albert Keiser, *The Indian in American Literature* (New York: Oxford University Press, 1933), passim; Williams, *Schoolcraft's Indian Legends,* passim.
2. James Fenimore Cooper, *The Last of the Mohicans* (New York: Grosset & Dunlap, n.d.), vi; Hodge, *Handbook of American Indians* 1:569; Beauchamp, *Aboriginal Place Names,* 238.
3. Romig, *Michigan Place Names,* 264.
4. Morgan, *League of the Iroquois* 1:64; idem, *Ancient Society* (Chicago: Charles H. Kerr Co., [1910]), 129–30, 132; Hodge, *Handbook of American Indians* 1:546; Schoolcraft, *Indian Tribes* 3:315.
5. Riggs, *Dakota-English,* 577.
6. Keating, *Narrative of an Expedition* 1:290–93; Mary Eastman, *Dahcotah, or Life and Legends of the Sioux* (New York: John Wiley, 1849), 165–73.
7. Romig, *Michigan Place Names,* 590–91.
8. Riggs, *Dakota-English,* 160–61, 314.
9. Baraga, *Otchipwe Language* 1:120.
10. Baraga has *kiwétin,* "the north wind, the wind going back." *Otchipwe Language* 1:299.
11. *Muskoday,* meadow or prairie, is *mashkodê* in Baraga, *Otchipwe Language* 1:168, 198; *nawadjiwan* is in ibid. 2:279.
12. Romig, *Michigan Place Names,* 590–91; Stewart, *American Place Names,* 344;

compare "Awake, I awake, *nin goshkos*" (etc.), in Baraga, *Otchipwe Language* 1:20.

13. Romig, *Michigan Place Names,* 421.
14. Compare Baraga: winter, *bibôn;* spring, *sigwan; Otchipwe Language* 1:290, 241.
15. Tanner, *Narrative of Captivity,* 396, 405.
16. Baraga, *Otchipwe Language* 1:214; Tanner, *Narrative of Captivity,* 312.
17. Baraga, *Otchipwe Language* 1:193; word for "pickerel" in Tanner, *Narrative of Captivity,* 311.
18. Baraga, *Otchipwe Language* 1:235.
19. Gagnieur, "Indian Place Names" (1918), 542; Romig, *Michigan Place Names,* 421. Baraga gives *opitchi* as the word for robin; *Otchipwe Language* 2:333; Tanner has *o-pe-che* in *Narrative of Captivity,* 305; whippoorwill is *waw-o-nais-sa* in ibid., 308.
20. Williams, *Schoolcraft's Indian Legends,* 155–59; Romig, *Michigan Place Names,* 313, 322; Lemoine, *Dictionnaire,* 264.
21. Romig, *Michigan Place Names,* 417; Fox, "Place Names of Cass County," 464.
22. Romig, *Michigan Place Names,* 514; Raudot in W. Vernon Kinietz, *The Indians of the Western Great Lakes 1615–1760* (Ann Arbor: University of Michigan Press, 1965), 380; Pierre F. X. de Charlevoix, *Journal of a Voyage to North America* (N.p.: Readex Microprints, 1966), 2:94; FWP, *Michigan,* 532–33; "Reports of Counties, Towns and Districts," *MPHSC* 1 (1874–75): 259–60.
23. Henry Whiting, *Sannilac, A Poem by Henry Whiting, with Notes by Lewis Cass and Henry R. Schoolcraft* (Boston: Carter, Hendee, and Babcock, 1831), 99; Osborn and Osborn, *Schoolcraft, Longfellow, Hiawatha,* 434, 447, 460, 527, 627, 647; "Reports of Counties," 315–16; Jenks, "County Names of Michigan" (1912), 454.
24. Keiser, *Indian in American Literature,* 75–77; Hodge, *Handbook of American Indians* 1:690–91; C. Henry Smith, Letter on the origin of Metamora, *Michigan History* 28 (April–June, 1944): 319–20; J. H. Trumbull, *Natick Dictionary,* Bureau of American Ethnology Bulletin no. 25 (Washington, D.C.: U.S. Government Printing Office, 1903), 56.
25. John Richardson, *Wacousta, or The Prophesy* (New York: Dewitt & Davenport, 1832); Louise Barnett, *Ignoble Savage* (Westport, Conn.: Greenwood Press, 1975), 10.
26. Keiser, *Indian in American Literature,* 249–52; Lowell Bean and Harry Lawton, *The Cahuila Indians of Southern California* (Banning, Calif.: Malki Museum Press, 1979), 11; Romig, *Michigan Place Names,* 465.
27. Lora T. Dickinson, *The Story of Winnetka* (Winnetka, Ill.: Winnetka Historical Society, 1956), 56–58; Vogel, *Indian Place Names,* 172–73; Baraga, *Otchipwe Language* 2:24, 417.
28. Stephen Powers, *Tribes of California* (reprint, Berkeley: University of California Press, 1976), 361.

# VIII
# Artificial "Indian" Names

The map of Michigan, more than that of any other state, is strewn with contrived half-Indian or pseudo-Indian names. Apparently the creators of these names, most notably Henry R. Schoolcraft, were striving toward euphony, but their products are unreal combinations, often joining syllables from two or more languages.

It is said that Benona, in Oceana County, is named for "a maiden in Indian legend," but details are lacking. The name could be a corruption of Winona, or it may be from some obscure nineteenth-century short story, novel, or poem. It is probably an artificial name, since it does not appear to be aboriginal.[1]

It has been indicated earlier that Michiana, in Berrien County, was named for its location on the Michigan and Indiana state line, and that Michillinda, in Muskegon County, was named by resort and cottage owners for the states of Michigan, Illinois, and Indiana. Nottowa, a village in St. Joseph County, takes its name from Nottowasepi (variously spelled), the name of a river and of a former Potawatomi reservation. *Nottowa* means "enemy," and *sepi* means "river." However, the name Wasepi, given to a village two miles north of Nottowa, is a meaningless fragment.

The name of the town and township of Zilwaukee in Saginaw County is an alteration of the name Milwaukee, for the Wisconsin city that was in its earliest years located within Michigan Territory. Supposedly Zilwaukee was named in 1854–55, in a conscious attempt to lure German settlers who had heard of Milwaukee. The name of the Wisconsin city means "good land," but Zilwaukee has no meaning.[2]

Most of the remaining artificial names indigenous to Michigan that are allegedly aboriginal appear to be the work of Henry R. Schoolcraft (1789–1865). Although born in New York State, he spent most of his life on Michigan's Indian frontier. He worked for nearly twenty years as agent to the Ojibwa at Sault Ste. Marie and Mackinaw, and he married Janet Johnston, a granddaughter of

Chief Wabojeeg. Negotiator of numerous Indian treaties, he was also a notable explorer, best known for his expedition to the source of the Mississippi in 1832. A prodigious writer, he turned out eight books and dozens of shorter publications. Aided by his patron, territorial governor Lewis Cass, who was Democratic candidate for president in 1848, Schoolcraft obtained financial support for his bulky, six-volume work ambitiously entitled *Information Respecting the History, Condition, and Prospects of the Indian Tribes of the United States* (1851–56).

He was a pioneer ethnologist, and his works are poorly organized and peppered with numerous errors. He has been charged with egotism, contradictions, and unreliability. One writer, Ackerknecht, has labeled him a "picturesque liar."[3] Schoolcraft was a member of the Michigan territorial legislature in 1828–32, when he introduced legislation creating a system of county and township names for the future state. So it was that many Indian, half-Indian, and pseudo-Indian names became ossified on the Michigan map, from which some of them spread to other states. It is due to Schoolcraft that Michigan has thirty-two county names that are wholly or partly drawn from Indian or alleged Indian names. Eighteen others were abolished after a short life, fifteen of them in a single day, March 18, 1843, in a sweeping revolt against such unfamiliar names as Notipokago. Schoolcraft, writing to the mayor of New York City on October 28, 1844, proposed that urban street names should be drawn either from aboriginal or American historical sources. In his *Indian Tribes* he published two papers on suggested geographical names for the American map.[4]

In the second paper he maintained that "the sonorousness and appropriate character of the Indian names has often been admired. They cast . . . a species of poetic drapery over our geography." He was no respecter of linguistic purity, and his proposed names consist largely of artificially joined syllables from unrelated Indian languages, names with Arabic prefixes, and suffixes from Latin, French, and English. Some are meaningless fragments of uncertain origin. Some of his worst inventions, fortunately, were rejected after a short life. Among the less objectionable of his names on the map are several that have puzzled name analysts.[5] Schoolcraft added to the confusion by giving inconsistent interpretations, in different places, of some of his manufactured names. Moreover, his frequent use of the prefix *al* is sometimes intended to represent the Arabic article, and at other times it stands for the first syllable of Algonquin.

Alcona is reportedly contracted from Arabic *al*, "the," *co* from an Ojibwa root meaning "plain" or "prairie," and *na* is held to mean "excellence." Alcona is the name of a county, village, township, and pond. Neither of the supposed Indian syllables in this name has the meaning assigned to them. *Co* is meaningless, and *na*, according to Baraga, indicates an interrogatory or is used as an interjection.[6] Here it is just a Latin suffix, as in India*na*.

Algansee, the name of a village and township in Branch County, is supposed to signify "Algonquin sea." It could also be composed of *Algan*, for Algonquin, plus the diminutive suffix *see*, thus meaning "Little Algonquin."

Algoma is a name that Schoolcraft unsuccessfully tried to fasten on Lake Superior. One of his more beautiful names, it was limited instead to a township in Kent County, where one story holds that the name was taken from that of a steamboat that plied the Grand River. Schoolcraft gave three variant explanations of this name: Algonquin plus *maig*, "waters"; "Lake of Algons," and *al*, from Algonquin, plus *goma*, "collected waters."[7] This name has been widely adopted outside Michigan. It is attached to a town and district of Ontario and to towns or villages in Mississippi, Oregon, Wisconsin, and West Virginia, as well as to a township in Minnesota. A spelling variant, Algona, became the name of villages in Iowa and Washington.

Algonac, the name of a town and state park in St. Clair County, was put together, according to Schoolcraft, from Algonquin and *ackee*, "land." Elsewhere he called it "Land of Algons."[8]

The name of Allegan County is an abbreviation of Allegany (also spelled Alleghany and Allegheny), the name of one of the two rivers that form the Ohio. Schoolcraft said it came from *Alligewi*, the name of a prehistoric Indian tribe. Allegheny (etc.) has also been called the Delaware equivalent of the Iroquois name Ohio, i.e., "beautiful river," and also explained as "river of the Allegewis."[9] Allegan is also the name of a town, township, and lake in Allegan County.

The village and township of Almena in Van Buren County have a name that Schoolcraft once defined as "an East Indian weight of about two pounds." There is also a claim that the name was given to the township by a state legislator, F. C. Annable, "after an Indian princess of whom he had heard." The red flag of caution should always be raised when an Indian princess is mentioned.[10] On its face, this name could have been composed of the Arabic article *al*, plus *men*, from Ojibwa *min* ("berry or fruit"), and a terminal *a* for euphony.

The name of the city and county of Alpena, Schoolcraft tells us, is composed of the Arabic article *al* and part of *penaisee*, Ojibwa for "bird." Elsewhere he said that *penai* means "partridge." In Baraga's dictionary and elsewhere this word is spelled *biné;* the shift from *p* to *b* has no significance. Just below Alpena is Partridge Point, an apparent English translation of Alpena.[11] The name Alpena has been borrowed in Arkansas and Nova Scotia.

Arenac County and Township have a name made from Latin *arena*, "sand," and Algonquian *akee*, "land," according to Schoolcraft.[12]

Iosco County has the name of an Ottawa hero in a legend published by Schoolcraft in his *Algic Researches* (1839). It is an artificial word that cannot be defined, yet spurious explanations are given. Schoolcraft himself said, "from *Iau*, to be; *os*, a father, and *coda*, a plain." However, *iau* does not appear in Baraga's dictionary, wherein *oôssima* is the word for "father," from which Schoolcraft extracted *os* without announcing the fragmentation. In like manner he drew *coda* from *mashkodê*, "prairie," without notice. Other ridiculous meanings in print are "water of light" and "shining water."[13] Iosco is the name of townships in Livingston County, Michigan, and Waseca County, Minnesota, as well as a lake in Passaic County, New Jersey. All were obviously taken from Schoolcraft's writings.

Oscoda County, containing a town and township with the same name, was established April 1, 1840. According to Schoolcraft, the name was taken "from *mushcoda*, a prairie or meadow, and *ossin*, a pebble." The explanation is correct, although the second element is more accurately translated as "stone."[14] In his combination, however, the name is meaningless.

The name of Oshtemo Township in Kalamazoo County is attributed to Schoolcraft, being interpreted as "head waters" and "head spring." It was apparently devised by extracting *Oshti* (altered to *Oshte*) from the Ojibwa word for "head," which is *oshtig-*

*wânima,* and adding the syllable *mo* from *mokidjiwanibig,* meaning "source, fountain, spring."[15]

Schoolcraft gave two irreconcilable explanations for the name Tuscola, which was given to a county and a township in 1840 and to a village in the county in 1875. In one place, he interpreted it as "warrior prairie" and in another as "from *dusinagon,* a level; and *cola,* lands." In neither case did he specify the language. All of his other made-up names in Michigan contain elements from Ojibwa or other Algonquian languages, but this one contains a borrowing from Choctaw *tushka,* "warrior." The meaning of *ola* is uncertain, but it cannot mean "prairie." *Ola* has been defined in Byington's Choctaw dictionary as a ring or sound made by a bell, but in this case it could be a corruption from *okla,* "people." More likely, however, it is a pseudo-Latin ending as seen in Indianola, Mineola, Neola, Wyola, and other names outside Michigan. The name Tuscola is also on the map in Illinois and Texas. Tuscola, Illinois, became a post office name in 1857. It is probable that the name in both Illinois and Texas was borrowed from Michigan.[16]

Some of Schoolcraft's artificial names are dealt with in our chapter on literary names. Because of their hybrid character, they are a mixed blessing, although definitely more appropriate than the classical names that deface New York's Mohawk Valley and other places. If something good can be said for these innovative names, it is in George R. Stewart's comment that they are "the most distinctive feature of American place-naming. Nowhere else has it so flourished."[17]

## Notes

1. Romig, *Michigan Place Names,* 56.
2. Ibid., 618–19.
3. Harlow Brooks, "The Medicine of the American Indians," *Journal of Laboratory and Clinical Medicine* 19 (October, 1933): 22; Erwin H. Ackerknecht, "White Indians . . . ," *Bulletin of the History of Medicine* 15 (January, 1944): 21.
4. Jenks, "County Names of Michigan" (1912), 439–78; Schoolcraft, "Plan of a System of Geographical Names," in *Indian Tribes* 3:501–8; and "Names Based on Indian Vocabularies . . ." in ibid. 5:621–25. His letter to the mayor of New York is appended as a note in the latter article.

5. Osborn and Osborn, *Schoolcraft, Longfellow, Hiawatha,* 357–65.
6. Ava Foster, "Indian Names," *Totem Pole* 29 (August, 1952): 2; Baraga, *Otchipwe Language* 2:260. Romig has "beautiful plain," in *Michigan Place Names,* 15. The only aboriginal element in this name is *co,* from Ojibwa *muscoda, mashkode,* etc., which is meaningless when separated.
7. Schoolcraft, *Indian Tribes* 3:509; 5:624; Romig, *Michigan Place Names,* 15–17; Osborn and Osborn, *Schoolcraft, Longfellow, Hiawatha,* 360.
8. Schoolcraft, *Indian Tribes* 3:509; 5:624.
9. Ibid. 5:536; George P. Donehoo, *Indian Villages and Place Names in Pennsylvania* (reprint, Baltimore: Gateway Press, 1977), 2–4.
10. Schoolcraft, *Indian Tribes* 5:593; Romig, *Michigan Place Names,* 19.
11. Schoolcraft, *Indian Tribes* 5:597; Baraga, *Otchipwe Language* 1:190.
12. Schoolcraft, *Indian Tribes* 3:536.
13. The Iosco story from *Algic Researches* is reprinted in Williams, *Schoolcraft's Indian Legends,* 139–47; see also Schoolcraft, *Indian Tribes* 5:624; Gannett, *American Names,* 165; Romig, *Michigan Place Names,* 285.
14. Schoolcraft, *Indian Tribes* 5:624; Baraga, *Otchipwe Language* 1:246.
15. Foster, "Indian Names," 5; Romig, *Michigan Place Names,* 421; Baraga, *Otchipwe Language* 1:238.
16. Schoolcraft, *Indian Tribes* 3:509; 5:624; Kane, *American Counties,* 357; Romig, *Michigan Place Names,* 563; "Reports of Counties," 322–23; Cyrus Byington, *A Dictionary of the Choctaw Language,* Bureau of American Ethnology Bulletin no. 46 (Washington, D.C.: U.S. Government Printing Office, 1915), 303, 601; Vogel, *Indian Place Names,* 153.
17. Stewart, *American Place Names,* xxxi.

# IX

# Rank, Gender, and Ethnic Names

The Ojibwa term for a child is given to Abinodji Falls in Presque Isle River, Gogebic County. The term *papoose* that is widely applied by whites to an Indian infant originated with the Narraganset of New England. Their vocabulary as given by Roger Williams includes: "Papòos, A childe." In Michigan are three Papoose lakes, in Iron, Kalkaska, and Lake counties.[1]

Chief is a widely used place-name in Michigan, in both English and aboriginal forms. In Manistee County are Chief Creek and Chief Lake, near which is the village of Chief. Another Chief Lake, besides Little Chief Lake, is in southwestern Marquette County. Father Verwyst listed but did not locate Chief's Mountain, a name translated from the Ojibwa *ogima,* "chief," and *wadjiw,* "mountain." The Ojibwa word for chief, in one of its spelling variants, appears in the name of Ogemaw Creek, Baraga County. In Gogebic County Ogima (Chief) Falls and Ogimakwe (Woman Chief) Falls are near each other in Copper Creek, a tributary of Presque Isle River. In the Lower Peninsula, as we have noted earlier, Ogemaw County is named for a particular chief, Ogema kegato. In the county are Ogemaw Creek, Ogemaw Lake, and Ogemaw Swamp. Arenac County has Ogemaw State Forest. Vanished examples of this name are the hamlet of Ogemaw in Iosco County, Ogemaw Springs in Ogemaw County, and Ogima, a mining settlement in Ontonagon County. Okemos (Little Chief), a suburb of Lansing, as earlier indicated, was named for an Ojibwa chief.[2]

The word *squaw,* for an Indian woman, originated in the Narraganset language of New England and was recorded in 1643 by Roger Williams.[3] Whites quickly applied it to Indian women of any tribe, and it became regarded, from misuse, as a derogatory term. Whites used Squaw as a place-name throughout the West and as far north as Alaska, but it is rare in New England and nonexistent in the South.

In Michigan the United States Geological Survey lists thirty-one place-names that include the word *squaw,* more than two-thirds of them being attached to creeks and lakes. In Michigan, as elsewhere, Squaw as a place-name usually arose from some episode of pioneer days, although the details are seldom recorded. One exception is the account of how Squaw Island in Orion Lake, Oakland County, was named. In 1826 a sawmill worker named Worden gave a jug of whiskey to some Indians, since he had no food for them. The women in the thirty-person party, anticipating trouble when the men became inebriated, disarmed them and went into the woods. After emptying the jug, the men demanded more and became threatening when their demands were unmet. Worden fled for his life. The next day he found that the women had taken refuge on an island, and so he called it Squaw Island.[4]

About the other *squaw* names in Michigan little or no information has been found. Squaw Creek in Berrien County is named "for an Indian woman who lived on the creek."[5] Another Squaw Creek is in Ingham County. Squaw Bay is in Lake Huron near Thunder Bay, Alpena County. Pioneers are said to have observed Indian women camping there. Squaw Island in Lake Michigan is a tiny bit of land north of Beaver Island, Charlevoix County. Squaw lakes are in Kalkaska and Marquette counties. Squaw Point is on the west side of Little Bay de Noc, Delta County.

The Shawnee word for "man" in the genderless sense, or more properly, "people," was *lenawai.* It is similar in other Algonquian languages, as seen in the name of the state of Illinois and the native name of the Delawares, Lenni Lenape. The name of Lenawee County, Michigan, however, mostly closely resembles the Shawnee form. Lenawee County was detached from Monroe County on November 20, 1826. Lenawee was once the name of a village in adjacent Hillsdale County. Lenawee Junction is a railway point in the present Lenawee County. In Bayfield County, Wisconsin, a lake is named Lenawee.[6]

Indian as a place-name occurs in every state in the union except Hawaii. At least seventy-seven Michigan place-names include the word. Additionally, a stream in Ontonagon County is called Native Creek. These names often mark present or former Indian village sites, and some topographic features are named for their supposed resemblance to Indian profiles.

Indiantown is the unofficial name of a small Potawatomi settlement near Athens, Calhoun County; Indiantown in Saginaw County is a former village site.[7] Indian Head, at Rock Harbor, Isle Royale, and Indianhead Mountain in Gogebic County are named for their appearance. Seven places are called Indian Point. Indian Point on Isle Royale is a prehistoric village site, occupied by Indians as late as 1847. Indian Point, the highest place on the north shore of Beaver Island in Lake Michigan, is reported to be a former Indian lookout site. Indian Drum Cave in the Pictured Rocks, Alger County, is U-shaped, and the breaking waves of Lake Superior cause a thundering, drumlike sound that can be heard for miles. One writer guessed that its name was inspired by *The Indian Drum,* a mystery tale of the Great Lakes written by W. B. MacHarg and Edwin Balmer (1917).

Twenty-eight water features in Michigan are called Indian Lake or Lakes. Indian Lake in Indian Lake State Park, Schoolcraft County, is fed by Indian River. Other Indian rivers are in Alger, Cheboygan, and Sanilac counties. Indian River in Cheboygan County gives its name to a village. Indian Fields Township, Tuscola County, recalls aboriginal cultivation at that place.[8]

Among other features are one each called Indian Bay, Brook, Channel, and Grove, two Indian islands, two Indian mounds, four Indian hills, and sixteen Indian creeks.

Indian tribes had different names for white people, and also for various European nationalities. To the Ojibwa, Ottawa, and Potawatomi, an Englishman was Saganash. In alternate spelling, it is seen in the name of Sogonosh Valley, Emmet County.[9]

According to several writers, the term *Yankee* was applied to the English by eastern Algonquians who had trouble pronouncing the word *English.* These writers held that some New England Indians pronounced it *Yengeese,* which evolved into Yankees. It was adopted by the Delawares but was not used in Virginia, where the Indians gave the English a name translated as "Long Knives."[10]

The name Yankee spread westward and became attached to a few place-names. In 1836 a former sheriff of Genesee County, New York, named William "Yankee" Lewis (1802–1853), settled on the Grand River in Barry County. He built a tavern that was an important stopping place for travelers. To feed them he grew "every" plant that the climate allowed. "Yankee" Lewis became a state legislator, and his establishment was called Yankee Springs, from its flowing water supply. Yankee Springs is today the name of the state recreation area at that site and of the township in which it is situated.[11]

In this era of color consciousness, it is interesting to notice that among the dozens of terms used by various Indian tribes to denote Europeans and various nationalities, very few indicated color. Pale face is an invention of white writers.[12]

## Notes

1. Baraga, *Otchipwe Language* 2:3; Roger Williams, *A Key into the Language of America* (reprint, Ann Arbor: Gryphon Press, 1971), 28.
2. Romig, *Michigan Place Names,* 413; Chrysostom Verwyst, "A Glossary of Chippewa Indian Names of Rivers, Lakes and Villages," *Acta et Dicta* 4 (July, 1916): 258.
3. Williams, *Language of America,* 27.
4. C. E. Carpenter, "Squaw Island—How it Received its Name," *MPHSC* 13 (1888): 486–88.
5. Fox, "Place Names of Berrien County," 10.
6. Mrs. Frank P. Dodge, "Landmarks of Lenawee County," *MPHSC* 38 (1912): 478; William A. Galloway, *Old Chillicothe, Shawnee and Pioneer History* (Xenia, Ohio: Buckeye Press, 1934), 316.
7. FWP, *Michigan,* 409; Romig, *Michigan Place Names,* 283.
8. Romig, *Michigan Place Names,* 283; Fred Dustin, "Isle Royale Place Names," *Michigan History* 30 (October-December, 1946): 704; FWP, *Michigan,* 147, 402, 409, 574, 607.
9. Nearly all writers spell this name with an initial *s* or *sh,* but Baraga renders it as *jâganash* in *Otchipwe Language* 2:88. In fact, he substitutes *j* for the *s* or *sh* of other writers in nearly every instance (*e.g.*, *jikag* for *shikag,* "polecat") probably because of his Slovenian origin. One of several guesses about the origin of *saganash* is that it represents an Indian attempt to pronounce the French *les Anglais*. For several spelling variations of this name see Vogel, *Indian Place Names,* 126 n. 815.
10. Heckewelder, *History of the Indian Nations,* 77, 142–43; Marquis de Chastellux, *Travels in North America* (Chapel Hill: University of North Carolina Press, 1963), 1:262n.; Gannett, *American Names,* 372. Some writers claim Dutch origin for Yankee, including Eric Partridge, *A Dictionary of Slang and Unconventional English* (New York: Macmillan, 1961), 968, and *Webster's New World Dictionary of the English Language,* 2d college ed. (n.p.: William Collins & World Publishing Co., 1978), 1646.
11. George H. White, "Yankee Lewis's Famous Hostelry in the Wilderness," *MPHSC* 26 (1894–95): 302–7.
12. See A. F. Chamberlain, "Race Names," in Hodge, *Handbook of American Indians* 2:348–53.

# X

# Material Culture Names

All names associated with Indian life ways are here designated as cultural names. The nonmaterial culture names—Manitou and its associated terms—are dealt with in the next chapter. Here we examine names relating to material culture, including names of dwellings, weapons, and implements; articles of clothing and adornment; names relating to hunting and fishing; and others.

The name of the city and county of Cheboygan probably means "big pipe." *Che* is the residue of Ojibwa *kitchi,* "big," while *boygan* is apparently a variant of *o-paw-gan, o-poi-gun, oppoygan,* or *opwagan,* and so on, "pipe." The reason for the name of this city at the eastern end of the straits of Mackinac is unknown, and there are several explanations of the name that differ from this one. One guess is that some topographic feature of the coastline had the appearance of a pipe.[1]

This name appears in Wisconsin as Sheboygan. There, too, its origin and meaning are in controversy. One folk tale has it that a local Indian chief had several daughters but no sons. When a new child was expected the chief hoped it would be a son, but another girl arrived instead. Announcing the event to his tribesmen, he said "She boy 'gain." The same story has been told of Cheboygan, Michigan.[2]

The name of the Cheboyganing River (or Creek), which joins the Saginaw River at Bay City, has the ending *-ing,* a locative suffix. Fred Dustin is undoubtedly correct in translating this name as "the place of the large pipe."[3]

Several names in Michigan are derived from Indian agricultural activity. In 1850, when lumbering and fishing attracted white men to the shore of Lake Michigan, in Delta County in the Upper Peninsula, they found a band of Menominee Indians cultivating fertile gardens, and for that reason they named the Garden Peninsula, the village of Garden, and Garden Township.[4]

Garden Island in Lake Michigan, Charlevoix County, is also

named because Indians planted gardens at that place. Indian Fields Township in Tuscola County was so named in 1852, as we have seen, because Indians raised corn and potatoes there.[5]

Ojibwa Indians on the shore of a lake astride the Michigan-Wisconsin border in Gogebic County once practiced farming. They called the place *Katekitiganing,* "the garden place" or "place of fields." The French, finding the "old deserted fields" of the Indians, called the lake *Lac Vieux Desert,* a name that survives today. A small band of Ojibwa Indians still resides on the north shore of the lake.[6]

A lake in Oakland County was, according to one writer, called *Menahasgorning,* "apple place," by the Indians, because of the apple trees and orchards nearby. (See Baraga's dictionary, where *mishiminatig* is defined as "apple tree.") The whites accordingly gave the name of Orchard Lake to that body of water and to the town established there in 1872.[7]

In Washtenaw County, wrote an early settler, was "a noted salt-lick of the red-deer and the red Indians." At one time this place was called the Salt Spring Reservation. Here the Indians extracted salt by boiling and evaporation. Consequently the whites gave the name Saline to the nearby river, as well as to a village and township near Ypsilanti.[8]

Indians also manufactured sugar and syrup from the sap of the maple tree, and they are generally credited with the invention of the techniques for producing those commodities. Many Sugar creeks and Sugar groves in the United States received their names because of the presence of Indian sugar camps. A large island in the St. Marys River near Sault Ste. Marie was called *Sisibakwato-miniss,* "Sugar Island," by the Ojibwa. The whites retained the name in English translation and gave it also to the township, in Chippewa County.[9]

Several Michigan names record aboriginal fishing and hunting customs. In 1670 Father Claude Allouez wrote of fish traps constructed by Indians on Wisconsin's Fox River.

> From one bank of the river to the other they make a barricade by driving down large stakes in two brasses of water, so that there is a kind of bridge over the stream for the fishermen, who, with the help of a small weir, easily catch the sturgeon

> and every other kind of fish, which this dam stops although the water does not cease to flow between the stakes. They call this contrivance Mitihikan, and it serves them during the spring and part of the summer.[10]

The aboriginal name of this fish weir, *mitihikan,* as given by Allouez, is equivalent to the word *mitchigan,* given by Baraga, which is translated "fence." In no way is it related to the state name. The Mitchigan River in the Upper Peninsula is named for this device. It is fed by the Fence River, the east branch of which originates in Fence Lake, Baraga County. The name Fence as applied to these water features is a translation from *mitchigan,* as is the name of Fence Lake in Vilas County, Wisconsin. The Fishdam River, a tributary of Big Bay de Noc in Delta County, has another name taken from that of the Indian fish weirs.[11]

Indians sometimes practiced netting or spearing fish at night, using pine knot torches for illumination. From this practice came the names of Torch Lake in Houghton County, Torch Lake in Antrim County, the village and township of Torch Lake along its shores, Torch River at the south end of the lake, and the village of Torch River in Grand Traverse County.[12]

Indians sometimes set fires to improve deer hunting by clearing the way for fresh growth that furnished deer browse. From that practice came the name of Burnt Plains in Baraga County. Gun Lake in Barry County is reported to have its name in translation from the native name, which was *Pashkisigan Sagaigan* in Ojibwa.[13]

The name of the most notable weapon of the Algonquians and other Indians has been adopted into English as *tomahawk*, from the language of the Powhatan (Renape) of Virginia. The word properly refers to a hatchet or axe, but it has been extended to war clubs. In Michigan, the name has been given to Big Tomahawk and Little Tomahawk lakes in Presque Isle County, besides Lower Tomahawk and Twin Tomahawk lakes in Montmorency County. It is not known whether the name in these places resulted from the discovery of tomahawks or from the shapes of the lakes. There are also Tomahawk creeks in Montmorency, Newaygo, and Presque Isle counties. The name Tomahawk is a white introduction, since it is not found in the native languages of Michigan. The Ojibwa word for "tomahawk" is *wagwadons.*

Snowshoes are an Indian invention and the inspiration for the name of Snowshoe Lake, Gogebic County. The Ojibwa name for "snowshoes" is *agim.*[14]

Of the many names of household utensils used by Indians, only one has been preserved on the map of Michigan: *Ontonagon.* In the Upper Peninsula it is the name of a county, town, township, river, and Indian reservation. The man-made places were named from the river. The county was established as *Ontonojan* on March 9, 1843, but the spelling was changed to Ontonagon five years later. Folk etymology relates that an Indian woman lost a dish in the river and shouted "*Nindonogan*" ("away goes my dish"), and so the stream was named. The Ojibwa word for "dish" is *onâgan,* which accounts for the name, except for the prefix *ont.* That is probably a variation of *-ond,* which Baraga says:

> in compositions, alludes to the *reason* or *origin* of [something]; to the *place* from which, or out of which, some object comes or is obtained.

Thus the name broadly means "where dishes are obtained." Perhaps some kind of clay or wood used in making dishes or bowls was obtained there. The Federal Writers Program translated this name as "place of the bowl." Other explanations in print, "hunting river" and "fishing river," have no factual foundation.[15]

Moccasin Shoals and Moccasin Bluff along St. Joseph River in Berrien County are said to be named from an Indian village that in turn was named for its chief. This was Potawatomi territory, but no such name can be found among those signed to the Potawatomi treaties. There are also Moccasin lakes in Alger, Gogebic, and Iron counties. *Moccasin,* as a word for the Indian shoe, has been naturalized into English from the languages of the eastern and central Algonquians. In Potawatomi, according to Gailland, it is *mokúsin.* Baraga spells it *makisin* in his Ojibwa dictionary. It varies only slightly in related languages.[16]

Of names of Indian dwellings, two are well known in English and are used as place-names. The Siouan word *teepee, tepee,* or *tipi* is not native to Michigan, but it is found in the names of Tepee Creek and Tepee Lake in Iron County.[17]

The word *wigwam,* and close variants, is the Algonquian equivalent term for any habitation. William Penn, writing of the

Delawares in 1685, said, "If an European comes to see them, or calls for lodging at their house or *Wigwam* they give him the best place and first cut."[18]

This word is given as *wigwam* in Potawatomi, by Gailland, *wig-wom* in Ottawa, by Blackbird, and *wigiwam* in Ojibwa, by Baraga.[19] The typical wigwam in this region was a dome-shaped dwelling made by a frame of saplings covered with bark or rushes. Some were oblong in shape. Wigwam Bay in Saginaw Bay, Arenac County, and Wigwam Picnic Area, Schoolcraft County, are apparently the only examples of this name on the Michigan map.

As we have seen, the Indian origin of many of our place-names is hidden because the names have been translated into English or French. The name of the Flint River, for which is named the city of Flint in Genesee County, is a translation of the Ojibwa name written as *Peonigoing sebe* (river of the place of flint) by Fred Dustin. More correctly, if we may rely on Baraga's dictionary, it should be *Biwânag sibi.* The river was named from the fact that Indians found chert, or flint, along its course, which was the main material for the manufacture of arrowheads and spearheads in the days before white traders wiped out that old craft with the introduction of metal arrowheads and guns.[20] The old craft of chipping arrow and spear points is suggested by the name of Chipping Creek in Ontonagon County.

Any place called Paint is probably so named because Indians obtained in that vicinity the materials for making paint that was used to adorn their bodies, costumes, and implements. Paint lakes and the Paint River, a tributary of the Brule River, all in Iron County, are examples of places named for that reason.[21]

The most popular paint among Indians was the mixture of yellow and red called *vermilion* by the French and *onaman, oulaman,* or *osanaman* by various Algonquian tribes. Longfellow recognized the importance of vermilion in his *Hiawatha* poem:

> Barred with streaks of red and yellow,
> Streaks of blue and bright vermilion,
> Shone the face of Pau-Puk-Keewis.

The village of Vermilion, on Lake Superior's shore in Chippewa County, is named for "red ochre deposits" (i.e., vermilion) that were found there.[22] Vermilion Creek, a tributary of the Looking Glass River in Clinton County, is another example of this group of names. As indicated elsewhere, the ancient names of Manistee, Manistique, and Nipigon contained prefixes (*Olaman*, *Onaman*, and so on) representing an Algonquian name for vermilion.

Pipes were indispensable items in Indian diplomacy. As Jonathan Carver relates, when Indian peace emissaries traveled to a council,

> They bear before them the Pipe of Peace, which I need not inform my readers is of the same nature as a Flag of Truce among Europeans, and is treated with the greatest respect and veneration, even by the most barbarous nations. I never heard of an instance wherein the bearers of this sacred badge of friendship were ever treated disrespectfully, or its rights violated. . . . It is used as an introduction to all treaties, and great ceremony attends the use of it on these occasions.

The bowls of Indian pipes were usually made of a red stone called pipestone by the whites, and today known as catlinite. The stems were of wood, and they were highly decorated. Places where pipestone was found were neutral areas. The notable Pipestone Quarry of Minnesota, now a national monument, is an example of such a place. In Berrien County, Michigan, Pipestone Creek was named because Indians obtained along its course the material for making pipes. The name was extended to a lake and a township.[23] (See also Calumet.)

The names of several Indian games are on the map in various states. The most popular Indian game in this region, *bagatiway*, called *la crosse* by the French, is recalled by the name of La Crosse Lake in Iron County. The name stemmed from the crosierlike racquet used in the game. A dozen states have places named La Crosse, but Snowsnake Mountain in Clare County is the only place in the country named for the game of snowsnake practiced by northern tribes. The game is played with a long, straight stick, often carved and painted to resemble an extended serpent. Each player hurls his dart along snow or ice or free in the air, to see whose missile will go the farthest. Stewart Culin says there were three main kinds of projectiles used, one a long polished red stick, another a bone slider with two feathers attached, and a third a javelin sometimes feathered and commonly tipped with horn. Culin gives this description, written in 1860, of the game among Ojibwa of the Apostle Islands in Lake Superior:

> The Indians are also said to have many capital games on the ice, and I had opportunity . . . to inspect the instruments employed in them, which they called Shoshiman (slipping sticks). These are elegantly carved and prepared, at the end they are slightly bent, like the iron of a skate, and form a heavy knob, while gradually tapering down in the handle. They cast these sticks with considerable skill over the smooth ice.

While the word *snowsnake* does not appear in this description, that is the common name by which the game became known among whites.[24]

*Wanigan* or *wanagan* is a lumberjack term borrowed from the Indians; it spread from coast to coast in the logging camps and also became a place-name. Its origin has been attributed to Abnaki *waniigan* or Cree *wunehi'kun,* "a trap," and to other sources. In woods parlance, it came to mean many things: a bag, box, or chest for food and supplies; a boat, rail car, or shack used as a cook house, commissary, or payhouse. The word was also used for the commissary or food wagon among sheepherders.

This interesting Americanism is preserved in the name of Wanagan Creek in Gogebic County, Michigan. There is also a Wannagan Creek in Billings County, North Dakota, and a small village called Winigan in Sullivan County, Missouri. In Hayward, Wisconsin, is a Wanigan restaurant.[25]

"In things relating to common life," wrote missionary David Zeisberger, "the language of the Indians is remarkably rich."[26]

## Notes

1. Pokagon, *Ogimawkwe Mitigwaki,* 157; Tanner, *Narrative of Captivity,* 225; Michel-Guillaume St. Jean de Crevecouer, *Journey into Northern Pennsylvania and the State of New York* (Ann Arbor: University of Michigan Press, 1964), 165; Baraga, *Otchipwe Language* 1:194; compare Verwyst, *jibaigan,* "any perforated object, as a pipe stem," in "Chippewa Names," 397. For other views, see

Romig, *Michigan Place Names,* 112; "Reports of Counties," 45–46; Kuhm, "Indian Place-Names," 112–13.

2. "Line o' Type or Two," *Chicago Tribune,* January 20, 1959; and Fred Dustin, "Some Indian Place-names around Saginaw," *Michigan History* 12 (October, 1928):734.
3. Dustin, "Some Indian Place-names," 734.
4. FWP, *Michigan,* 541.
5. Romig, *Michigan Place Names,* 283.
6. William F. Gagnieur, "Ketekitiganing (Lac Vieux Desert)," *Michigan History* 12 (October, 1928): 776–77; the United States Geological Survey calls this place Katakitekon.
7. Romig, *Michigan Place Names,* 419. The last part of the aboriginal name is corrupted, for the sound of *r* is not in the languages of the Three Fires; Baraga, *Otchipwe Language* 1:15; compare Potawatomi *michi-mini-ki-wo-ke-ku,* "orchard," in Gailland, "English-Potawatomi," 229.
8. L. D. Norris, "History of Washtenaw County," *MPHSC* 1 (1874–76): 328; FWP, *Michigan,* 390.
9. C. A. Browne, "The Chemical Industries of the American Aborigines," *Isis* 23 (1935): 408; Charlotte Hamilton, "Chippewa County Place Names," *Michigan History* 27 (October–December, 1943): 639; Gagnieur, "Indian Place Names" (1919), 417.
10. Allouez, in Kellogg, *Early Narratives,* 150.
11. Baraga, *Otchipwe Language* 1:99; 2:254; Verwyst, "Chippewa Names," 259.
12. Romig, *Michigan Place Names,* 558; FWP, *Michigan,* 521.
13. FWP, *Michigan,* 567; Baraga, *Otchipwe Language* 1:122, 153; this name was garbled into Par-ke-gon-bish in Weissert, "Indians of Barry County," 323.
14. William H. Holmes, "The Tomahawk," *American Anthropologist,* n.s., 10, (1908): 264–76; Baraga, *Otchipwe Language* 1:236, 265.
15. "Reports of Counties," 307; Gannett, *American Names,* 232; Baraga, *Otchipwe Language* 1:74; 2:327; FWP, *Michigan,* 598.
16. Romig, *Michigan Place Names,* 374; William R. Gerard, "Virginia's Indian Contribution to English," *American Anthropologist,* n.s., 9 (1907): 97; Gailland, "English-Potawatomi," 325; Baraga, *Otchipwe Language* 1:227.
17. *Típi,* "a tent, house, dwelling, abode," Riggs, *Dakota-English,* 470.
18. Albert C. Myers, ed., *Narratives of Early Pennsylvania, West Jersey and Delaware* (reprint, New York: Barnes & Noble, 1959), 232.
19. Gailland, "English-Potawatomi," 160; Blackbird, *History of the Ottawa and Chippewa,* 123; Baraga, *Otchipwe Language* 1:136.
20. Baraga, *Otchipwe Language* 1:105, 214; Dustin, "Some Indian Place-names," 732; Gannett, *American Names,* 127.
21. Kelsie Harder, *Illustrated Dictionary of Place Names* (New York: Van Nostrand Reinhold, 1976), 408.
22. Baraga, *Otchipwe Language* 1:278; Hamilton, "Chippewa County Place Names," 642.
23. Carver, *Travels,* 358–59; Fox, "Place Names of Berrien County," 25.
24. Stewart Culin, *Games of the North American Indians* (reprint, New York: Dover Publications, 1975) (reprinted from *Twenty-Fourth Annual Report, Bu-*

*reau of American Ethnology* [Washington, D.C.: U.S. Government Printing Office, 1907], 399–400, citing J. G. Kohl, *Kitchi-Gami* [London: 1860], 90).

25. Mitford W. Mathews, *Dictionary of Americanisms* (Chicago: University of Chicago Press, 1956), 1829; E. A. Watkins, *A Dictionary of the Cree Language* (Toronto: Anglican Book Centre, 1981), 207; Gerard in Hodge, *Handbook of American Indians* 2:910; Alexander F. Chamberlain, "Algonkian Words in American English. . . . ," *Journal of American Folk-Lore* 15 (October–December, 1902): 265; Walter F. McCulloch, *Woods Words: A Comprehensive Dictionary of Logging Terms* (Corvallis: Oregon Historical Society, 1977), 207; Ramon Adams, *Western Words: A Dictionary of the American West* (Norman: University of Oklahoma Press, 1968), 340, 354.
26. David Zeisberger, "A History of the Indians," *Ohio Archaeological and Historical Quarterly* 19 (January & April, 1910): 144.

# XI
# The Spirit World

Places named Manitou frequently have some legend connected to them. From the legend of Sleeping Bear Dunes, as indicated elsewhere, came the names of North and South Manitou islands in Lake Michigan. They represent the two cubs that drowned while trying to swim across the lake behind their mother.

The name of Montowibo Creek, a tributary of the Black River in Gogebic County, is probably a corruption of Ojibwa *manito-wabo,* literally "spirit liquid," or *manito-waba* (in Cree, *munito-wupaw*), "strait of the spirit." The latter is the general explanation given by Baraga and others for the name of Manitoba, in Canada. Another similar name is Manitowabing Lake, in the Parry Sound district of Ontario, in which *-ing* is a termination signifying "place." A chief named Manitowaba, of White River, in lower Michigan, is mentioned in a treaty signed at Washington, March 28, 1836, but these places are not named for him.

There are villages called Manitou Beach in Lenawee and Presque Isle counties. The first is located on Devil's Lake, which has a village of the same name on its shores. The story behind this name is that an Indian girl, daughter of the Potawatomi chief Meateau, was drowned in the lake, and her body was not recovered. The chief presumed that she was carried off by evil spirits.[1]

The Algonquian tribes of this region called all spirits, good or bad, *manitou.* The great or good spirit was *kitchi-manitou,* a bad spirit was *matchi-manitou,* and ordinary spirits were simply *manitou.* Countrywide, it is not uncommon to find the name Manitou in association with "Devil" in place-names, for, to many whites, all Indian deities were devils. The United States Geological Survey lists twenty-two Michigan place-names containing the word Devil or Devils, and it appears likely that most of them represent a white translation of Manitou.

Manitou churned up massive storms in Lake Superior, and the rocky point of the Keweenaw Peninsula became a graveyard of canoes and, later, ships. The island off the point is called Manitou Island, and at each end of it a navigation light warns ships of its hazards. A variant spelling of Manitou is found in Manido Falls, in the Presque Isle River, Ontonagon County. It is an Anglo spelling compared to the French form, Manitou. Another variant is Manito, on a lake in Oakland County. Manitou, however, is most frequent. Besides the places already mentioned, it appears on Manitou Lake in Shiawassee County, Lake Manitou in Leelanau County, and Manitou Payment Highbanks and Manitou Point in Mackinac County.

The Dead River, which joins Lake Superior at Marquette, has a waterfall clouded in mists. The Indians believed that these mists carried the spirits of deceased Indians into the spirit world, and so the river was called *gi-nibo-manitou-sibi,* or some variant thereof, meaning "river of the spirits of the dead." The French called it *Riviére de Mort,* which the Americans translated into Dead River.[2]

Anything powerful, mysterious, or sacred the Indians called by a name that we have translated as "medicine." Although to us the term implies specific remedies or treatments, the aboriginal culture also applied it to an entire complex of rituals, sacred objects, and secret societies. The Midéwiwin was the grand medicine society among the Algonquians of the northern lakes.

In many western states the term *medicine* has been attached to certain topographic features that were believed to have certain spiritual associations. Nearly always the name is on the map in English, but in Michigan is an exception, Mahskeekee Lake, in Delta County. The name is a reasonable representation of the Ojibwa term for "medicine," recorded as *mashkiki* in Father Baraga's *Otchipwe Language.* In our language, this place would be called "Sacred Lake."[3]

An Indian magician, doctor, or sorcerer was called a *wabeno,* "dawn man," by the Ojibwa and Potawatomi Indians. Members of the fourth degree of the Midéwiwin, or medicine society, were wabenos. They made hunting medicine, love powders, and magic cures. The yarrow plant, which had certain medical and ritualistic uses, was called *wabeno-wusk.* The title of the honored practitioners of the occult arts is preserved in the name of Wabeno Creek, a tributary of the Little Carp River in Ontonagon County. A town and township in Forest County, Wisconsin, are also named Wabeno.[4] White men regarded wabenos and other Indian seers as witches, and so, very probably had them in mind when they named

Witch Lake in Marquette County as well as Magician Lake in Cass County.

Among some northeastern Algonquians a personage similar to our lakes area wabeno was called a *powow* or *pow-wow.* Roger Williams listed "Powwow: A Priest" in his Narraganset vocabulary of 1643. Its original meaning, "priest or conjuror," came from a root referring to "dreamer." From there the meaning and applications of the term evolved into both verb and noun, having to do with curing sessions, talking, meetings, rituals, dancing, and the like. It was adopted especially by Pennsylvania Germans to describe their Indian-like curing procedures, which involved a mixture of herbalism and magic or spiritual exercises. Today the term *powwow* is used by Indians from coast to coast to describe social or religious gatherings for dancing, contests, and athletic events. Whites use the term as a word for meetings, caucuses, and boisterous events. Today it is an adopted foreign word for both Indians and whites. It is also a place-name in Rockingham County, New York, Washakie County, Wyoming, and Gogebic County, Michigan, which has Powwow Lake.[5]

According to Ojibwa legend, a tribe of giant cannibals lived on an island in Hudson Bay. They were called *windigos* and were greatly feared by all the Indians about the northern lakes. In *Algic Researches,* Schoolcraft reproduces one tale of a "ferocious Weendego," told by the Saginaw band of Ojibwa. These creatures also appear in the *Hiawatha* story, where Mudjekeewis exhorts Hiawatha:

> Cleanse the earth from all that harms it,
> Clear the fishing-grounds and rivers,
> Slay all monsters and magicians,
> All the giants, the Wendigoes. . .

At times, Windigo (Windego, or Weendego) was used as a personal name. Wabi Windego (White Giant) was an Ottawa Indian of Grand River who signed a treaty at Washington on March 28, 1836.

As a place-name in Michigan, Windigo appears as the name of a place on the west end of Isle Royale in Lake Superior, which is the site of Windigo Inn and the dock for boats from Grand Portage, Minnesota. The name at this place was taken directly from a copper mine that was operated there from 1890 to 1892 by the Wendigo Copper Company. In another alternate spelling, we also have Lake Windaga in Isabella County.[6]

"In the world-view provided by Ojibwa religion and magic," wrote anthropologist Ruth Landes, "there is neither stick nor stone that is not animate."[7] Michigan's place-names are the richer for it.

## Notes

1. Baraga, *Otchipwe Language* 1:299; 2:5; Watkins, *Dictionary of the Cree,* 186, 344; Kappler, *Indian Treaties,* 455; FWP, *Michigan,* 532–33; Romig, *Michigan Place Names,* 155.
2. Romig, *Michigan Place Names,* 148, 155; Pokagon, *Ogimawkwe Mitigwaki,* 122; Baraga, *Otchipwe Language* 1:68; Tanner gives *Nebowesebe* for "Dead River," in *Narrative of Captivity,* 168; Baraga has it as in my text; Romig offers *Djibis-manitou-sibi* in *Michigan Place Names,* 148; Ruth Landes lists *djibai* as "a ghost or spirit other than manito," in *Ojibwa Religion and the Midewiwin* (Madison: University of Wisconsin Press, 1968), 239; Gagnieur, confusing Dead River with Carp River, called it Namebini-sibi, "Sucker River," in "Indian Place Names" (1918), 540.
3. Virgil J. Vogel, *American Indian Medicine* (Norman: University of Oklahoma Press, 1970), 25; Baraga, *Otchipwe Language* 2:223; Landes, *Ojibwa Religion,* passim.
4. Walter J. Hoffman, "The Midewiwin or 'Grand Medicine Society' of the Ojibwa," *Seventh Annual Report, Bureau of American Ethnology* (Washington, D.C.: U.S. Government Printing Office, 1891):159; Vogel, *American Indian Medicine,* 22; Kuhm, "Indian Place-Names," 124.
5. Chamberlain, "Algonkian Words," 71–72; Williams, *Language of America,* 127; Hodge, *Handbook of American Indians* 2:303; Mathews, *Dictionary of Americanisms,* 1297–98.
6. Williams, *Schoolcraft's Indian Legends,* 169–74; Hodge, *Handbook of American Indians* 2:930; Baraga, *Otchipwe Language* 1:41; Kappler, *Indian Treaties,* 454–55; Dustin, "Isle Royale Place Names," 722.
7. Landes, *Ojibwa Religion,* 21.

# XII

# Names from Fauna

Indian tribes sometimes took their names, or had their names given, from animals—e.g., Erie, "wildcat." Tribes were divided into clans that were named for animals, birds, plants, or natural phenomena—e.g., thunder. These tutelary spirits were represented by the Ojibwa word *totem.* Many personal names were also taken from the same sources, and so too were names of places. (See Peshekee ["Buffalo"] River, Marquette County.)

Places named for animals, birds, reptiles, and other fauna were sometimes so named because of topographical features supposed to resemble them. Some were named for legendary or actual events involving the animal. Some were named for the presence or frequency of the animal in the area. For that reason, these names frequently record the former range of certain animals. Some of the other places were named by whites, who may have brought the name from another region.

The importance of the beaver to the Ojibwa is shown by the fact that Baraga lists twenty-two words relating to the animal, its age or condition, body parts, dams, and lodges. *Ahmeek* (*amik*) is the general Ojibwa name for "beaver," which animals were once abundant in the Upper Peninsula. Whites gave the name to the village of Ahmeek in Keweenaw County, which was incorporated in 1909, to Ahmeek Lake on Isle Royale, and Amik Lake in Gogebic County. Ahmikwan Lake in Lake County recalls beaver lodges once located there. Beaver Island in Lake Michigan was called by the Ojibwa *Amikokenda,* "home of the beavers." Miramichi Lake in Osceola County has a name that, according to Bishop Baraga, arose from the Cree word *mayamisk,* "ugly beaver."[1]

A large member of the deer family formerly found in the old Northwest was called elk by Europeans, but *kalibu* (scratcher, pawer) by the Micmac Indians in the Northeast and in Canada's maritime region. The name referred to the animal's habit of pawing beneath the snow for forage in winter. The name passed into French as *caribou* and from thence to English.[2]

In Michigan the caribou were last seen on Isle Royale in Lake Superior, from which they disappeared about 1912, to be replaced by the moose. Their name survives on Caribou Creek, Isle Royale, and the offshore islands called Lower and Upper Caribou. Another Caribou Creek joins Lake Superior in Ontonagon County. The former presence of the animal is also recorded in Chippewa County by the names of Caribou Creek and Caribou Lake, with a village of the same name on its shore. The name caribou was introduced into Michigan by whites, for according to Baraga, the Ojibwa did not distinguish this animal from the moose.

The name of the moose is appropriately found on Moose Lake and Moose Point, Isle Royale, which is the only place in Michigan where this largest member of the deer family still roams. The former range of the animal is indicated by the names of Moose Lake in Alger, Dickinson, and Iron counties, Moose Lakes in Luce County, and Moosehead Lake in Gogebic County. Mosinee Creek in Gogebic County has its name from the plural word for "moose." The name *moose* is of eastern Algonquian origin and signifies "eater," for the animal's browsing habit.[3]

Bears had human qualities in the Indian view, and the animal was prominently situated in their spirit world and legends. Bears still roam the wilder parts of Michigan, and eighty-three places, mostly creeks and lakes, are named for them. However, the aboriginal name is on only one feature, Muckwa Creek in Mason and Lake counties. It is a reasonable approximation of Ojibwa *makwâ.*[4]

John Smith of Virginia wrote in 1612 that "There is a beast they call *Aroughcun,* much like a badger, but useth to live on trees as squirrels do." From that Powhatan name evolved the word *raccoon,* which was popularly shortened to *coon.* Raccoon or Coon is a popular name for creeks or topographic features throughout those regions where this nocturnal animal is numerous. Although they are found in all forty-eight of the lower continental United States, they are less common in the North than in the South, and this is reflected in the distribution of the name. In Michigan, there is a Coon Lake in Baraga, Grand Traverse, Lake, and Livingston counties. Coon Creek is a tributary of the Clinton River in Macomb County, while other Coon creeks are in Ingham and Shiawassee counties. Moreover, there is Coon Hill in Jackson County, Coonskin Creek in Oceana County, and Coontail Lake in Baraga County. Formerly, when the coon was the symbol of the Whig

Party, the name Coon Town was given to a village in Lenawee County, because of the numerous Whigs there. Today it is called Addison.[5]

The name of Waugoshance Island and Point, in Wilderness State Park, Emmet County, has been translated as "Little Fox." *Wau-goosh* is the Potawatomi-Ottawa word for "fox," but the last part of the name, *ance,* is conceivably from French *anse,* a "bay" or "cove." Alexander Henry, the English trader captured at Mackinac in 1763, called the place Wagoshense and translated it "Fox Point." According to Ivan Walton, lake sailors once called this place Wobble Shanks.[6]

The Fox Islands in Lake Michigan, according to Schoolcraft, were called *Annamosing,* "place of the little dog," by the Ojibwa and Ottawa Indians. "It is," he wrote, "simply the diminutive and local forms united, added to the elementary term for a dog." Another writer contended, however, that the proper name was Waugooshe-minis, "Fox Island."[7]

Hog Island in Lake Michigan, northeast of Beaver Island, Charlevoix County, was called *Kokoshi minissing,* "Hog Island place," by the Ojibwa.

Gogomain is the name of a former village in Chippewa County; it is preserved on a river and swamp in that county. The name is reputed to mean "porcupine," which is given as *Kâg* by Baraga and *Kahg* by Tanner. It could be a perversion of *kâgwaiân,* Chippewa for "porcupine skin."

The dark eminence of the Porcupine Mountains along Lake Superior's shore in Ontonagon County was first described by the geographer Jedidiah Morse, who reported in 1823:

> On the east side of this river [Presque Isle] are the *Porcupine Mountains,* which are shaped like this animal, from which they take their name, extending nine miles along the shore of the Lake, rising to the height of about eleven hundred feet, sloping toward the shore.

His estimate of elevation was conservative, for Government Peak, the highest elevation of this range, rises 1,248 feet above the lake. Thomas L. McKenney, on July 24, 1826, noted the "prodigious elevation" of these mountains but gave no figure. Henry R. Schoolcraft, on July 11, 1831, gave the Chippewa Indian name of the range as *Kaug Wudju,* and upon inquiry to Chief Konteka of Ontonagon

River he was told that they were so named because of their resemblance to a crouching porcupine. Father Verwyst (1916) gave the Chippewa name of this place as *Kagwadjiw*, "Porcupine Mountain."[8]

The Ojibwa called the skunk *jikâg* (Baraga) or *she-gagh* (Tanner). The common name of the striped beast, however, descends from one of the aboriginal dialects of New England. As a place name, skunk is not exceedingly popular, but Michigan has a Skunk Lake in Van Buren County and Skunk creeks in Dickinson, Isabella, Mackinaw, and Ogemaw counties. It is possible that the Ojibwa term for skunk is represented in the names of Chicago Lake in Delta County and Chicagon Lake and Slough in Iron County. However, because of the possibility of floral origin, these names are discussed in our chapter on floral names.[9]

Wolves are extinct in Michigan, except on Isle Royale, although there has been an unsuccessful attempt to reintroduce them in the Upper Peninsula. A toponymic memorial to the storied predator is the name of Maingan Lake, Gogebic County. It has the Ojibwa name for "wolf."

The animal commonly known today as the groundhog or woodchuck was called by the name of another animal, the fisher, among some Indians and French Canadians. The Ojibwa name for that animal was *otchig*, and from that evolved our word *woodchuck*.[10] Michigan has a Woodchuck Creek in Monroe County.

Meguzee Point on Grand Traverse Bay in Antrim County could have its name from *migisi*, Ojibwa for "eagle." An Ottawa chief of L'Arbre Croche, who was mentioned in a treaty signed at Washington on March 20, 1836, was named Megisawba, but no data have been found connecting him to this place.

Partridge Point on Lake Huron below Alpena has the English translation of the Ojibwa name *pena*, to which Schoolcraft prefixed the Arabic article *al* to form *Alpena*. The name of Penasa Lake in Oscoda County means "little partridge."[11]

A small lake on Isle Royale is called *Shesheeb*, Ojibwa for "duck." A lake near Negaunee, in Marquette County, was called *Shishibegomad*, "floating ducks," by the Ojibwa.[12]

Insect names are not common as place-names, but a few do exist. Watassa Lake in Mason County almost surely has a name cut from *wawatessi*, defined as "glow worm," but more often linked with "firefly" or "lightning bug." (Compare Wah-wah-taysee in *Hiawatha*, and Wauwatosa, Wisconsin.)[13]

Places named for fish outnumber all other native names for fauna on the Michigan map. They are a testimonial to the importance of fish in the aboriginal economy. The Carp River (one of several streams having this name), which joins Lake Superior at Marquette, was called Namebini-sibi by the Ojibwa. The Indians had the same name for the sucker, but whites translated the name as carp, for a fish introduced from Europe. The old Ojibwa name, in the plural, survives on Namebinag Creek in Schoolcraft County.[14]

Siskowit, or cisco, is the common name given to several species of fish found in the Great Lakes and surrounding waters. Lake herring, lake mooneye, lake trout, and fresh water salmon have all been called by this Ojibwa name, which is said to be abbreviated from *penitewiskowet,* "that which has oily flesh."[15]

Both Siskowit (or Siskiwit) and Cisco are well represented on the map of Michigan as well as northern Wisconsin. Cisco Bayou in White River, Oceana County, is perhaps the only example in the Lower Peninsula. Cisco Lake is in Gogebic County, and Cisco Branch joins the south branch of the Ontonagon River in Ontonagon County. Another Cisco Lake is in Bayfield County, Wisconsin, along with Siskiwit Lake, from which the Siskiwit River flows to Siskiwit Bay of Lake Superior. On Michigan's Isle Royale are Mt. Siskiwit, Siskiwit Bay, Siskiwit Falls, Siskiwit Lake, the Big and Little Siskiwit rivers, and Little Siskiwit Island. The names on Isle Royale apparently stem from that of a copper mining company that operated there from 1845 to 1855.[16]

In Ojibwa and Potawatomi, *gigo* is a generic term for "fish." Longfellow spelled it *keego* in *Hiawatha,* and it is so spelled in place names: Keego Harbor in Oakland County, Michigan, and the village of Keego in Ontario. There are also Kego Lake and Township in Minnesota. Kegomic, a village in Michigan's Emmet County, may have its name from a corruption of *gigoika-magak,* "abundance of fish."[17]

In Alger County, the Laughing Whitefish River flows from Laughing Whitefish Lake, roars over a cataract called Laughing Whitefish Falls, and discharges into Lake Superior at Laughing Whitefish Point, midway between Marquette and Munising. This landmark was frequently mentioned by early travelers. According to Gagnieur, Laughing Whitefish is a translation from the Ojibwa name, *Atikameg Bapit.*[18]

A large fish of the northern lakes was called *maskinonge,* "ugly fish," by the Ojibwa. Baraga said "it may be a kind of Jackfish, who has a peculiar bump on his back." The Ojibwa name was used by the French, from whom it passed into English. Alexander Henry, the English trader, wrote that "Among the pike, is to be included the species called by the Indians, *masquinong.*" The old spelling, *maskinonge,* is still found on some Canadian lakes, and the scientific name of the fish became *Esox masquinongy.* Gradually the common name became *muscallonge, muskellunge, muskellonge,* and so on, and was widely used as a name for lakes and rivers in the upper Great Lakes region. Michigan has Muskellonge Lake and state park in Luce County, Muskellonge Bay in Mackinac County, and Muskellunge Lakes in Montcalm and Montmorency counties.[19]

The largest fish in the Great Lakes is the sturgeon. The Ojibwa name for it was *namê* or *nahma,* and Mishi-Nahma, "King of fishes," appears in *Hiawatha.* In Delta County in the Upper Peninsula, the name appears in those of the village and township of Nahma, as well as Nahma Junction. In English translation it appears as the name of the Sturgeon River, which joins Big Bay de Noc at Nahma. The Ojibwa name of the river is Namesibi. Nimikon Falls in the Presque Isle River, Ontonagon County, is a name that appears to mean "place where there are sturgeon" (*namê,* sturgeon, + *ikan*).[20]

The preceding name seems to be related to that of Naomikong Point on Whitefish Bay, a name also on a creek, lake, and island, all in Chippewa County, along Lake Superior. Naomikong is said to be the old Indian name of Whitefish Point in Luce County, twenty miles to the north. Two explanations for it are in print, both of which seem to be wide of the mark. Gagnieur said it was derived from *onimik* ("sprout, bud") and *neashing* ("point") and that it was given because of the abundance of sprouts or buds there. Schoolcraft believed it meant a place of abundant beaver (*na,* "abundance," *amik,* "beaver," *ong,* "place"). That is a kind of arbitrary carpentry that Schoolcraft condemned in others. From the circumstances of the place, it appears more credible to hold that Naomi is probably nothing but a variant of *namê* or *nahma,* meaning "sturgeon," but used at times as a generic name for "fish," and *-ikong,* a variant of *-ikan,* "place where." The best free translation appears to be "fishing place."[21]

Ogontz Bay in Delta County is "Bay of Okans or little herring," according to Gagnieur. The name is also on a river and locality in Delta County. Baraga has *okiwiss* for "herring," while *okans* is "young pickerel." Larry Matrious, a Hannahville Potawatomi, explained Ogontz as "little pickerel." An Ottawa Indian named Ogonse signed a treaty at Fort Industry, Ohio, on July 4, 1805. Some places are named for him elsewhere, but he lived too far from Delta County to be the source of the name of Ogontz Bay and River, which are clearly named for the pickerel.[22]

Reptilian place-names are scarce. Kenabeek Creek ("Snake Creek"), as previously suggested, probably has its name from *Hiawatha.* Mackinac, Mackinaw, and Michilimackinac may be the most famous reptilian names in Michigan and in the country, if in fact they signify "turtle," or, with the prefix *michili,* "great turtle." The meaning of this name, however, is in sharp dispute, and so is its pronunciation. Part of the confusion may result from the variety of recorded spellings of this name, sixty-eight being listed in one source.[23]

Mackinac County is on the north side of the Straits of Mackinac, while Mackinaw is the name of a lake in Chippewa County and of the city on the south side of the straits. It is the site of the original Fort Michilimackinac, which the French built in 1715 and surrendered to the British in 1760. In 1779, the British, fearing American attacks, built a new fort on Mackinac Island. That fort was turned over to the Americans in 1796, retaken by the British in 1812, restored to the Americans in 1815, and closed in 1842. Both forts are now restored and part of the state park system.

Mackinac is pronounced *Mackinaw* by Michigan residents, but some outsiders have remarked that "this may well startle the owners of Cad-in-laws and Pon-ti-aws." One of the earliest definitions of Mackinac, and still the most generally accepted, is that given in the Raudot memoir (ca. 1710):

> The Outavois [Ottawas] live at the post of Michilimakina. . . . an island opposite gave it its name of Michilimakina, which means the turtle, because it seems to have the shape of this animal, which is very common there.[24]

The "turtle" definition was also set forth by Lamothe Cadillac, Pierre Charlevoix, Jonathan Carver, Alexander Henry, Juliette Kinzie, Jedidiah Morse, John Tanner, Bela Hubbard, Bishop Baraga, and Chrysostom Verwyst, among others.[25] However, there are other stories. Bacqueville de la Potherie (1753) gave this account of the origin of "Michilimakinak":

> Michilimakinak, according to the old men, is the place where Michapous [Great hare, a spirit] adjourned longest. There is a mountain on the shore of the lake which has the shape of a hare; they believed that this was the place of his abode, and they call this mountain Michapous. . . . There is an island, two leagues from shore, which is very lofty, they say that he left there some spirits, whom they call *Imakinagos.* As the inhabitants of this island are large and strong, this island has taken its name from those spirits; and it is called *Micha-Imakinak*—for in the Outaöuak language *micha* means "great," "stout," and "much."[26]

Alexander Henry reaffirmed the earlier view that Mackinac signified "A Turtle" and Michilimackinac "is the Great Turtle," so named from the resemblance of the island to that creature. He added that the Algonquins used the *l* sound in the name, but the Chippewa (Ojibwa) substituted *n.*[27]

Andrew J. Blackbird, an educated Ottawa, maintained (1887) that Mackinac was named for an ancient tribe allied with the Ottawa—the Mi-shi-ne-macki-naw-go, from whence the place was called Mi-shi-ne-macki-nong. These Indians, he wrote, were massacred by the Seneca in the days before whites arrived. However, his contemporary, Simon Pokagon, a Potawatomi, derived the name from *michi mi-ki-nock,* "a big snapping turtle."[28]

Later writers have given other interpretations, none of which seem convincing. They include "place of the big wounded or lame person," by William Jones, a Fox Indian (1907), and "land of the great fault," by Emerson R. Smith (1958).[29] These citations do not exhaust the storehouse of views on this name, but the others have still less credibility. It is this writer's own view that the weight of evidence sustains those who maintain that Michilimackinac means "great turtle" and Mackinac means "turtle." That conclusion is supported not only by most accounts of early visitors (La Potherie excepted), but also by linguistic sources. In Ojibwa, Baraga has *makina* and *mikkina* for "turtle"; Watkins gives the Cree word for "turtle" as *mikinak;* Lemoine has *mishtinak* for "great turtle" in Montagnais. We endorse the view of A. F. Chamberlain (1902):

> The place-name *Mackinac* (*Mackinaw*) would represent an [in?] Ojibwa (or closely related dialect) *makinâk* ("turtle"), but the

word is said to be really a shortened form of *Michilimackinâc*, a corruption of *mitchi makinâc* ("big turtle").[30]

Aside from the controversial Mackinac, Michigan has Mishike Lake in Gogebic County. Its name, according to Bishop Baraga, refers to "a kind of large tortoise." The turtle is deeply embedded in aboriginal lore. According to the *Walum Olum*, the traditional history of the Delawares, the people and animals took refuge on the back of a great turtle when the evil manitou sent the deluge.[31]

## Notes

1. Baraga, *Otchipwe Language* 1:28, 300; Romig, *Michigan Place Names,* 13; Dustin, "Isle Royale Place Names," 694; Gagnieur, "Indian Place Names" (1919), 412.
2. Alexander F. Chamberlain, "Significance of Certain Algonquian Animal Names," *American Anthropologist,* n.s., 3 (1901): 678.
3. Dustin, "Isle Royale Place Names," 697, 710; Baraga, *Otchipwe Language* 1:123; Chamberlain, "Algonkian Words," 249; Verwyst, "Geographical Names," 394.
4. Baraga, *Otchipwe Language* 1:24.
5. Lyon G. Tyler, ed., *Narratives of Early Virginia 1606–1628* (reprint, New York: Barnes & Noble, 1959), 93; Romig, *Michigan Place Names,* 132.
6. Haines, *American Indian,* 792; Pokagon, *Ogimawkwe Mitigwaki,* 129; Armour, *Attack at Michilimackinac,* 63; Ivan H. Walton, "Indian Place Names in Michigan," *Midwest Folklore* 5 (Spring, 1955): 34.
7. Schoolcraft, *Indian Tribes* 3:527; compare *animosh,* "dog," in Baraga, *Otchipwe Language* 1:79; William W. Johnson, "Indian Names in the County of Mackinac," *MPHSC* 12 (1897): 380.
8. Romig, *Michigan Place Names,* 227; Baraga, *Otchipwe Language* 1:197; Tanner, *Narrative of Captivity,* 303; Verwyst, "Chippewa Names," 261, 268; Jedidiah Morse, *Report to the Secretary of War on Indian Affairs* (New Haven: S. Converse, 1822), appendix, p. 29; McKenney, *Sketches,* 212; Schoolcraft, *Personal Memoirs,* 360.
9. Hodge, *Handbook of American Indians* 2:596; Virgil J. Vogel, *Iowa Place Names of Indian Origin* (Iowa City: University of Iowa Press, 1983), 83–85; Baraga, *Otchipwe Language* 1:196 (Polecat); Tanner, *Narrative of Captivity,* 303.
10. Baraga, *Otchipwe Language* 1:291; Chamberlain, "Algonkian Words," 267.
11. Baraga, *Otchipwe Language* 1:83; Kappler, *Indian Treaties,* 455; Schoolcraft, *Indian Tribes* 5:267.
12. Dustin, "Isle Royale Place Names," 716; Gagnieur, "Indian Place Names" (1918), 542.
13. Baraga, *Otchipwe Language* 2:404.
14. Ibid., 269.

15. Ibid. 1:42, 249; Hodge, *Handbook of American Indians* 1:300; 2:580; Mathews, *Dictionary of Americanisms,* 1550; Verwyst, "Geographical Names," 297.
16. Kuhm, "Indian Place-Names," 115; Romig, *Michigan Place Names,* 516; Dustin, "Isle Royale Place Names," 707, 717.
17. Baraga, *Otchipwe Language* 1:103; Pokagon, *Ogimawkwe Mitigwaki,* 54; compare *kegoi* in Schoolcraft, *Indian Tribes* 6:676; Baraga, *Otchipwe Language* 2:131.
18. Gagnieur, "Indian Place Names" (1918), 539.
19. Baraga, *Otchipwe Language* 1:299; Chamberlain, "Algonkian Words," 247; Alexander Henry, *Travels and Adventures* (Ann Arbor: University Microfilms, 1966), 30.
20. Baraga, *Otchipwe Language* 1:248, 270; Tanner, *Narrative of Captivity,* 310; A. S. Gatschet, "The Fish in Local Onomatology," *American Anthropologist* 5 (October, 1892): 361–62; Kelton, *Indian Names of Places,* p. 43; Romig, *Michigan Place Names,* p. 287; Verwyst, "Geographical Names," 394; Gagnieur, "Indian Place Names" (1918), 550.
21. Gagnieur, "Indian Place Names" (1919), 418; Schoolcraft, *Indian Tribes* 3:506; 5:593; Baraga, *Otchipwe Language* 1:248, 270.
22. Gagnieur, "Indian Place Names" (1918), 550; Baraga, *Otchipwe Language* 2:316, 321; Larry Matrious, interview, August 22, 1984; Kappler, *Indian Treaties,* 78.
23. Hodge, *Handbook of American Indians* 1:857.
24. Kinietz, *Indians of the Western Great Lakes,* 379; this document is also called, in some sources, the DeLiette memoir and DeGannes memoir.
25. Cadillac memoir in Milo M. Quaife, *The Western Country in the Seventeenth Century* (New York: Citadel Press, 1962), 3; Charlevoix wrote: "The name of Michillimakinac signifies a great quantity of turtles, but I have never heard that more of them are found here at this day than elsewhere." *Journal* 2:46; Carver, *Travels,* 19; Juliette Kinzie, *Wau-Bun, or the 'Early Day' in the Northwest* (Chicago: Rand McNally, 1901), 25; Morse, *Report,* appendix, p. 607; Tanner, *Narrative of Captivity,* 304; Bernard C. Peters, ed., *Lake Superior Journal: Bela Hubbard's Account of the 1840 Houghton Expedition* (Marquette: Northern Michigan University Press, 1983), 78; Baraga, *Otchipwe Language* 1:300; Verwyst, "Geographical Names," 392.
26. Blair, *Indian Tribes* 1:287. Other legends of Mackinac are in Dirk Gringhuis, *Lore of the Great Turtle* (Mackinac Island, Mich.: Mackinac Island State Park Commission, 1970).
27. Henry, *Travels and Adventures,* 37, 107n., 110. He refers to the Algonquin *tribe,* not the Algonquian family of tribes.
28. Blackbird, *History of the Ottawa and Chippewa,* 19–20; Pokagon, *Ogimawkwe Mitigwaki,* 131.
29. Hodge, *Handbook of American Indians* 1:857; Emerson R. Smith, "Michilimackinac, Land of the Great Fault," *Michigan History* 42 (December, 1958): 392–95; see also George S. May, "The Meaning and Pronunciation of Michilimackinac," *Michigan History* 42 (December, 1958): 385–90.
30. Baraga, *Otchipwe Language* 1:299; Lemoine, *Dictionnaire,* 280; Watkins, *Dictionary of the Cree,* 210; Chamberlain, "Algonkian Words," 246–47.
31. Baraga, *Otchipwe Language* 2:248; Dunn, *True Indian Stories,* 188.

# XIII
# Names from Flora

Seventy years ago the botanist Melvin R. Gilmore remarked on ways in which vegetation helps shape the culture of a region. Remarking that "the prevalence of certain plants often gave origin to place names," he cited former Indian names of Nebraska streams that were derived from the local flora. His views can be confirmed in Michigan.[1]

Michigan has a Chicago Lake in Delta County and a Chicagon Lake in Iron County. Neither is named for the great city at the foot of Lake Michigan, but the names of all three could be etymologically related.

There are two principal explanations of Chicago, one being that it refers to the wild leek, onion, or garlic; the other that it signifies skunk in one or more Algonquian languages. Since the aboriginal terms for both the plants and the animal have a common root, the meaning of the place-names derived from them can best be settled by reference to historic documents and the environment.

The meaning of the name of the city of Chicago seems to be settled beyond a reasonable doubt by the testimony of Henri Joutel, a survivor of LaSalle's Texas expedition, who wrote in his journal of September 26, 1687, that "Chicagou" took its name "from the quantity of garlic which grows in this district, in the woods."[2]

That interpretation, which was surely obtained from the Indians, was reiterated by LaMothe Cadillac (1718), who wrote that the word *Chicago* "means Garlic River, because a very large quantity of garlic grows wild there, without cultivation."[3] August Chouteau wrote (1816) of "Garlick Creek, or the real Chicago on Lake Michigan."[4] Others who identified the name of Chicago, Illinois, with the wild garlic, leek, or onion (the last called "skunk weeds" by Tanner), are John Tanner and Henry R. Schoolcraft.[5] So far as we can discover, the first to link the city's name to the striped animal, or anything else, was Mrs. John H. Kinzie, who asserted (1855):

> The origin of the name Chicago is a subject of discussion, some of the Indians deriving it from the fitch or polecat, others from the wild onion with which the woods formerly abounded; but all agree that the place received the name from an old chief who was drowned in the stream in former times.[6]

In this region, Chief Checagou of the Illinois is the only historic chief of that name; he was a visitor to the French Court in 1725. Since the place-name was in use by whites as early as 1682, in LaSalle's accounts, the chief could not have been the protonym. He was never a resident of Chicago; moreover, naming places for individuals was contrary to custom among unacculturated Indians.

Baraga's dictionary has the following:

> CHICAGO (Cree), from *chicâg,* or *sikâg,* a skunk, a kind of wild cat, word, which at the local term, makes chicâgôk.

In Ojibwa, he offered *jîgawanj* for "garlic." The initial syllable, *jig,* is *shig* in Tanner, while the Potawatomi Simon Pokagon gave She-gog-ong as the city's name; thus we have another example of the difficulties of faithfully recording aboriginal sounds.[7]

Since the origins of the names of Michigan's Chicago and Chicagon lakes are obscure, we lack historic documentation of their meaning. If named for the "smelly plants," the names originally would have possessed the terminal syllable *wanj* (Baraga) or *winje* (Tanner). However, terminal sounds are frequently dropped in white usage. Since both skunks and the odoriferous plants are found in the Upper Peninsula, these lakes could be named for either.

The village of Copemish in Manistee County has a name so mangled that it is difficult to analyze. One writer interprets it as "beech tree." The Ojibwa term for "beech tree" is given as *ajawêmij* by Baraga and *uz-zhuh-way-mish* by Tanner. Copemish could have been extracted from one of the following terms, from which one or more initial syllables were dropped: *wa-go-be-mish,* "bark tree," (Tanner); *sis-se-go-be-mish,* "willow," (Tanner); or *wigobimish,* "basswood" (Baraga). The last seems most probable, because of the importance of basswood for the carving of wooden implements and the use of its bark for weaving material. (See also Wico and Bois Blanc.)

Gijik Creek, Gogebic County, has the uncorrupted Ojibwa name for the cedar tree, and is probably named for the white cedar, or Arbor vitae *(Thuja occidentalis)*.[8]

The hickory tree received its name from the Powhatan Indians of Virginia. However, their term from which hickory was derived was applied not to the tree but to a milky gruel made from the nuts of this and other trees. John Smith reported that various nuts were dried, beaten to a powder, and mixed with water. The mixture was "coloured as milke; which they cal *Pawcohiscora,* and keepe it for their use." A. F. Chamberlain, authority on Algonquian words in English, held that, "From the cluster words *pawcohiccora,* etc., transferred by the whites from the food to the tree, has been derived *hickory.* The latter form was in use by 1682."[9]

Of the several species of hickory, the commonest in this region is Shagbark hickory (*Carya ovata*). Hickory trees are commonly found only in the Lower Peninsula of Michigan, and the state has far fewer places named for them than are found in the lower Midwest states. These include Hickory Corners, a village in Barry County, Hickory Creek in Berrien County, Hickory Hills, a ski area near Traverse City, Hickory Island in Wayne County, Hickory Lake in Shiawassee County, and Hickory Ridge, a settlement in Oakland County.

Although pine forests, particularly of white pine (*Pinus strobus*) once contributed mightily to the economy of Michigan, only a single place-name records the aboriginal (Ojibwa) name of the genus *Pinus:* Jingwak Lake in Gogebic County.[10]

In the fall of 1830, Juliette Kinzie embarked from Detroit with her new husband, John H. Kinzie, to travel via Mackinac to his post as Indian agent at the site of present Portage, Wisconsin, which was then in Michigan Territory. At their new home, she was upset by the smoking habits of their Indian visitors: "I watched the falling of the ashes from their long pipes," she complained, "and other inconveniences of the use of tobacco, or Kin-ni-kin-nick, with absolute dismay."

Kinnickinick, variously spelled, was a smoking mixture that sometimes contained real tobacco but was always mixed with leaves of sumac (*Rhus glabra*) or bearberry (*Arctostaphylos uva-ursi*) and the bark of red willow (*Cornus sericea*). The name of this substance, derived from Ojibwa *kinikinige,* "he mixes," became as widespread as the use of the substance. White trappers and woodsmen, especially the French, used it as commonly as the Indians. It is not surprising that its rhythmic name became a place-name in seven states and one Canadian province. In Gogebic County, Michigan, is Kinnikinnick Creek (so spelled).[11]

The sugar maple tree (*Acer saccharum* Marsh), one of several species called sugar maple, is found throughout Michigan and is especially abundant in the Upper Peninsula. It is the leaf of this tree that became the national emblem of Canada. It afforded the sap used by Indians and pioneers for the manufacture of syrup and sugar, causing it to be commemorated by innumerable Sugar creeks, groves, and the like. The name of Manakiki Falls on Maple Creek in Gogebic County retains an Ojibwa Indian name meaning "forest of maple trees." Probable derivatives of this name are Manoka Lake in Montcalm County, Manake Lake in Mecosta County, and Lake Manuka in Otsego County.[12]

Of all the wild plants growing in Michigan, none was more important to the aboriginal population than wild rice, the *manomin* of the Ojibwa (*Zizania aquatica*). As a major food source, it sustained denser populations than the exclusively hunting and fishing regions farther north. Albert O. Jenks, in his notable monograph on this plant at the beginning of this century, listed 160 geographic names derived from it, in both English and aboriginal languages. He concluded that more geographic names were derived from this plant "than from any other natural vegetal product throughout the entire continent."

For reasons unknown, few of these names are found in Michigan. As we have seen, the city, township, and county of Menominee were named from the Menominee River, which took its name from the Menominee tribe, which in turn was named for its principal food source. The only other Michigan names arising from this plant, as listed by Jenks, are in English: Rice Lake in Newaygo County, Rice Creek, a tributary of the Kalamazoo River at Marshall, and its source, Rice Lake, in Calhoun County.[13]

Mitigwaki Creek and Lake in Iron County have as a name the generic Ojibwa term for a "forest."

A wide expanse of St. Marys channel below Neebish Island in Chippewa County is called Munuscong Lake. The Munuscong and Little Munuscong rivers discharge into it. Along the shores of the lake are Munuscong State Forest and the village of Munuscong. Writers of this century have held that this name means "bay of the rushes." Schoolcraft wrote of it: "Min-us-co, from minno, *good,* and *ushk,* an aquatic plant."[14]

The *ushk* mentioned by Schoolcraft is apparently akin to *ashkin,* which Baraga defined as "something raw." This term, somewhat mangled, is seen in the name of the Ush-kab-wan River, in St. Louis County, Minnesota. That name has also been written Ushkabwahka and translated "wild artichokes."[15] The term *min,* which has been converted to *mun* in our lake name, sometimes occurs in other spellings and is often prefixed to the names of edible plants—for example, *menomin,* "rice," and *mondamin,* "maize." Therefore, the name may not signify rushes as commonly supposed. Although rush stems are edible, their primary use was for lodge coverings and other purposes. Baraga's word for "rush mats," *anakanashk,* lacks this prefix, but the name for rushes is seen in the syllable *ashk.*[16]

The name Paw Paw, which occurs in four counties of southwestern Michigan, confused Father Verwyst, who called it the Chippewa term for "papa," or "father," just as whites used the words Pa and Papa.[17] There can be no question, however, that as a place-name, Paw Paw refers to a shrubby tree (*Asimina triloba*).[18] Its common name is ultimately from the Haitian aborigines, while the generic part of the scientific name, *Asimina,* is from the Algonquian, most probably the Illinois.[19] The present common name is a corruption of *papaya,* from the tropical fruit. It was described by Oviedo in 1526 but not named until later, when it was called by the Taino term *ababai.*[20]

The *papaw* or *pawpaw* of North America, which also produces an edible fruit, was so named for its fancied resemblance to the Caribbean fruit. "The Papau," wrote John Lawson (1714), "bears an Apple about the bigness of a Hen's-Egg, yellow, soft, and as sweet as any thing can well be."[21]

Wherever pawpaw grew, it was prized as food by the pioneers. Its range in Michigan is confined to the southern part of the Lower Peninsula. The name is usually spelled as one word in books, but as two words on the Michigan map. In Berrien County is Paw Paw Lake, which gave its name to a town upon its shores, and the Paw Paw River. Other Paw Paw lakes are in Hillsdale and Kalamazoo counties, while the town and township of Paw Paw are in Van Buren County.

The Pinconning River, a tributary of Saginaw Bay, gave its

name to the town of Pinconning, in Bay County. From the Ojibwa language, this name is interpreted as "potato place." Its root, as given by Baraga and Tanner, is *opin,* "potato."[22] Undoubtedly it is not named for the cultivated potato, but for one of the wild tuberous plants sometimes called wild potato, such as the Jerusalem artichoke (*Helianthus tuberosus*) or the arrowhead (*Sagittaria latifolia*).

The Pinnebog River and the village of Pinnebog in Huron County are said to receive their name from the Ojibwa name for "partridge drum." The name appears, however, to be simply a slight alteration of *binébag,* "partridge leaf." The substitution of *b* for *p* has no significance, as these letters are interchangeable. The name probably refers to the partridge berry (*Mitchella repens*).[23]

Michigan has at least fourteen lakes named *Tamarack,* besides Little Tamarack, and another, in Eaton County, miscalled Tamarock. There is also a Tamarack River in Gogebic and Iron counties, a Tamarack Swamp in Ogemaw County, three Tamarack creeks, and a Tamarack community in Houghton County. The name Tamarack is a corruption of a northeastern Algonquian name for the American larch (*Larix* species), of which there are three species. The tree is an evergreen that flourishes in bogs. Bartlett's *Dictionary of Americanisms* (1860) gave *hackmatack* as the original form of this word. A. F. Chamberlain said that the name was "generally thought to be derived from some of the Algonkian dialects of Canada or the New England States," possibly from *ackmatuck* or *ackmestuk,* signifying "wood for bows and arrows."[24]

Wapato Creek, a tributary of Lake Gogebic in Ontonagon County, has a name reported to signify "white root," i.e., potato, in Cree or Ojibwa. The reference is not to cultivated potatoes, but to the bulbous, edible roots of the arrowhead plant (*Sagittaria* species). Its English name describes the shape of its leaves. Growing in streams and marshes, its bulbs were harvested and roasted by the Indians for food. The word *wapato* traveled west with the Indian and French-Canadian trappers, and was early adopted into the Chinook jargon of the Northwest. Lewis and Clark, arriving on the Columbia River in 1805, received some of these roots from the Indians, and from them they named Wappatoo Valley and an island near the mouth of the Willamette. Wapato remains a town name on the Yakima reservation in Washington.

The closest approximation to Wapato in Baraga's *Otchipwe*

dictionary is *wâbado*, "rhubarb," but it is conceivable that Wapato could be a mangled form of *watapin*, "a small edible root." In Watkins's Cree dictionary the term most similar to *wapato* is *wapitāo*, "it is faded white."[25]

The village and township of Waucedah, in Dickinson County in the Upper Peninsula, have a name that is almost certainly of Winnebago origin and is perhaps a transfer name from that of a former village in Adams County, Wisconsin. The differing views are Romig's, that it is "Indian for over there," and Verwyst's, holding that it is from Ojibwa *wassiti*, "it reflects light."[26]

However, the name is more probably a Winnebago term meaning "at the pines" or "place of pines," from the root *wa-zee* or *wazi*.[27] Besides the now vanished name in Adams County, Wisconsin, Waucedah is clearly related to the name of Wauzeka, in Crawford County, Wisconsin, which has been interpreted as "pine" or "large pine," and Wayzata, in Hennepin County, Minnesota, which has been explained as Dakota-Siouan for "pines," "at the pines," and "north."[28] Winnebago is a language of the Siouan family.

Birch bark had many uses among northern tribes—for covering lodges and canoes and for making *makoks* (boxes) and numerous implements. The sap was used for medicine in respiratory ailments. Wegwaas Lake in Chippewa County is named from the generic Ojibwa name of the "birch tree," *wigwâss*. It is obvious that Lake Wequos, in Otsego County, has its name slightly altered from the same source.[29]

Michigan has thirty-nine place-names containing the European word *birch* but only two having the native name for that beautiful tree.

## Notes

1. Melvin R. Gilmore, *Uses of Plants by the Indians of the Missouri River Region* (reprint, Lincoln: University of Nebraska Press, 1977), 4–5.
2. Vogel, *Indian Place Names*, 69.
3. Quaife, *Western Country*, 69.
4. August Chouteau, "Notes on the Indians of North America," MS from *Ancient and Miscellaneous Surveys* 4, St. Louis, February 2, 1816 (National Archives), 3.
5. Tanner, *Narrative of Captivity*, 298; Schoolcraft, *Indian Tribes* 5:573.
6. Kinzie, *Wau-Bun*, 150.

7. Baraga, *Otchipwe Language* 1:298; 2:170; Tanner, *Narrative of Captivity,* 298; Pokagon, *Ogimawkwe Mitigwaki,* 141.
8. Gannett, *American Names,* 91; Baraga, *Otchipwe Language* 1:23, 25; 2:132; Tanner, *Narrative of Captivity,* 294.
9. Smith, in Tyler, *Narratives,* 91; Hodge, *Handbook of American Indians* 1:547.
10. Baraga, *Otchipwe Language* 1:193.
11. Kinzie, *Wau-Bun,* 76; Hodge, *Handbook of American Indians* 1:692; Caleb Atwater, *The Indians of the Northwest, Their Manners, Customs, &c. or Remarks made on a Tour to Prairie du Chien. . . .* (Columbus: 1850), 104.
12. Baraga, *Otchipwe Language* 2:216.
13. Jenks, "Wild Rice Gatherers," 1115–26.
14. Baraga, *Otchipwe Language* 2:255; Hamilton, "Chippewa County Place Names," 639; Romig, *Michigan Place Names,* 385; Schoolcraft, *Indian Tribes* 5:624.
15. Warren Upham, *Minnesota Geographic Names* (reprint, St. Paul: Minnesota Historical Society, 1969), 499.
16. Baraga, *Otchipwe Language* 1:142, 217.
17. Verwyst, "Geographical Names," 296.
18. Fox, "Place Names of Berrien County," 8; Romig, *Michigan Place Names,* 492.
19. William R. Gerard, "Plant Names of Indian Origin," *Garden and Forest* 9 (July 15, 1896): 283.
20. Gonzalo Fernandez de Oviedo, *Natural History of the West Indies* (reprint, Chapel Hill: University of North Carolina Press, 1959), 25; current dictionaries list this plant as *papaw,* with *paw-paw* as an alternate spelling.
21. John Lawson, *History of North Carolina* (Richmond: Garrett & Massie, 1937), 107.
22. Romig, *Michigan Place Names,* 442; FWP, *Michigan,* 488; Tanner, *Narrative of Captivity,* 298; Baraga, *Otchipwe Language* 1:197.
23. Romig, *Michigan Place Names,* 445; Baraga, *Otchipwe Language* 1:190.
24. John R. Bartlett, *Dictionary of Americanisms: A Glossary of Words and Phrases . . . Peculiar to the United States,* 3d ed. (Boston: Little Brown, 1860), 186; Chamberlain, "Algonkian Words," 244, 260.
25. Mathews, *Dictionary of Americanisms,* 1829; Chamberlain, "Algonkian Words," 265; Baraga, *Otchipwe Language* 2:390; Watkins, *Dictionary of the Cree,* 501; George Gibbs, *A Dictionary of the Chinook Jargon or Trade Language of Oregon* (New York: Cramoisy Press, 1863), 28; Elliott Coues, ed., *The History of the Lewis and Clark Expedition* (reprint, New York: Dover Publications, n.d.), 2:693.
26. Romig, *Michigan Place Names,* 586; Verwyst, "Geographical Names," 397.
27. Norton W. Jipson, "Winnebago Vocabulary" (Chicago Historical Society, 1924? MS), 389; Paul Radin, *The Winnebago Tribe* (reprint, Lincoln: University of Nebraska Press, 1970), 67.
28. Kuhm, "Indian Place-Names," 138; Robert E. Gard and L. G. Sorden, *The Romance of Wisconsin Place Names* (New York: October House, 1968), 136; Upham, *Minnesota Geographic Names,* 207; Riggs, *Dakota-English,* 50, 563.
29. Baraga, *Otchipwe Language* 2:414.

# XIV

# Rivers, Streams, and Lakes

A cursory look at the map of Michigan soon reveals an abundance of aboriginal names on water features, with French names close behind. Both have shown considerable resistance to change. Europeans found the Indian names of lakes and streams in place and so recorded them as they found them. Sometimes the French simply translated them into their own language (Ecorse, for example). Less commonly, they substituted their own names, as in the case of Sault Ste. Marie (*Pawating,* "the rapids," to the Ojibwa), now the rapids and river of St. Marys. The English and Americans made some changes, including some translations into their own language, but the preponderance of existing names for the larger lakes and streams held fast. When they named their settlements, it was a different story. Finding no names to uproot, in most instances, they could establish their own names, which frequently duplicated those of their old homes in the East or in Europe.

However, we must not exaggerate the Indian influence on water names. Michigan is reported to have five thousand lakes. Of course the Indians did not have names for all of them, and often the names they did have were not recorded or did not survive. Even today some small lakes are unnamed. Settlers were able to supply their own names to the smaller lakes and streams, while names of many larger features remained aboriginal. Unfortunately, most of the water feature names in Michigan are dreary repetitions of such common names as Bass Lake (48), Mud Lake (210), and Silver Creek (40). There are also a number of "Indian" names given by whites, such as Squaw Creek or Lake (21), in addition to names from literature or legend (Hiawatha), names from back East, indigenous Indian names bestowed by whites, and the undetermined number of names translated from Indian languages.

The Bad River, a tributary of the Saginaw River at Saginaw, has a name that is translated from the Ojibwa name, Matchi-sibi.

As one writer wrote poetically:

> The river that's called Mich-a-see-be,
> Bad river it is, by English name.

A "bad" river usually was difficult for canoe travel due to rapids or obstructions. Schoolcraft called this river "Mauvaise or Maskigo River," conveying the French word for "bad" and the Ojibwa word for "swampy."[1]

The name of Chikaming Township in Berrien County was given in 1842 by Richard Peckham, from a Potawatomi name meaning "big lake." The original form was probably Kitchi-gaming, "at the big lake." Since there are no large lakes in the township, the name probably refers to Lake Michigan, which it borders.[2]

The town of Dowagiac in Cass County is named for Dowagiac Creek and the Dowagiac River, into which the creek flows. Authors have explained this name as signifying "I am going hunting fish," "scoop up," "fishing river," and "foraging ground." None of these finds support in vocabularies of the Potawatomi who inhabited the area.

Mr. James Clifton mentions a Potawatomi Indian named Dwagek (Autumn) who was tried and acquitted in connection with the massacre at Indian Creek, in Illinois, May 21, 1832. The residence of this Indian is not known, but a corruption of the Potawatomi word for autumn could have been adopted by whites as a name for these features. The word for autumn in Gailland's Potawatomi dictionary is *tekwakuk.* The equivalent in Ojibwa is *tagwâgi* (fall), or *tagwâgig* (in fall). The only other Dowagiac in the United States is in Hart County, Kentucky. There it is undoubtedly a transfer name.[3]

Lake Gogebic in Gogebic and Ontonagon counties has a name that is difficult to explain, because it is obviously much altered from its original form. Some writers hold that it was originally *Agogebic,* but there is disagreement about the meaning of that as well. Explanations of the name in that form include "body of water hanging on high," from its location on an elevation (Verwyst), "trembling ground" (Romig, citing Havighurst), and "rocky or rocky shore" (Haines). Verwyst also offered the alternative form *gogibic,* "diving place."

In the absence of known historical evidence, we must turn to Baraga's *Otchipwe Language.* The nearest approach to the present name to be found there is "Rock, *âjibik,* on the rock, *ogidâbik.*" The meaning of the name cannot, apparently, be determined with any certainty. The name was adopted by the Gogebic Mining Company (1853), apparently from the lake. Two post offices named for it, Gogebic and Lake Gogebic, were closed many years ago. Gogebic State Park is on the lake's southwestern shore.[4]

The winding Grand River is Michigan's longest river. Passing through seven counties in its 185-mile journey, it begins in Jackson County, flows through the cities of Lansing and Grand Rapids, and empties into Lake Michigan at Grand Haven.

The Ojibwa name for the Grand River, according to Baraga, was *Washtenong.* However, he did not explain it. The name was altered to Washtenaw (which is today a street name in Ann Arbor) and given to the county of Michigan Territory proclaimed by Governor Lewis Cass on September 10, 1822, and organized on November 29, 1826. One explanation of the name (1876) is that it means "at or on the river," which is clearly erroneous. Another view (1881) maintains that the original form of the name was "*Wuste-nong* or *Washte-nong,* meaning literally the 'further district' or 'land beyond,' further country." In 1838, Schoolcraft mentioned *Washtenong* as the name of a sternwheel steamer operating on Grand River. In 1905 Dwight Goss called the Grand River "O-wash-ta-nong, or the far-away-water, so named because it was the longest river in the territory." The most aberrant explanation is from Verwyst (1916): "Washtano-sibi, Washington river, from Washtenong, to or from Washington."[5]

The phrase "far away" appears repeatedly in most explanations of this name. That could not be correct, however, if the initial sound was O. Ojibwa *wassa,* "far," is the source of the name of Wausau, Wisconsin.[6] It seems likely, however, that O was not a part of the original name, for it is not in the earlier documents.

The aboriginal name of the Grand River may be related to that of Detroit, which Baraga gave as Wawiiatan, from *wawiieia,* "it is round or circular," which is apparently compounded with *atan,* "town." With the addition of the suffix *ong,* often seen in references, the name signifies, according to Baraga, "in or at Detroit, to or from Detroit." The city name referred to the curved channel of the Detroit River. Schoolcraft wrote the name of Detroit as *Waweeatunong,* formed from *wa-wea,* defined as "a maze or circle, winding to all points of the compass," and *tun,* from *ah-tun,* "a channel," and the locative *ong.* However, Baraga's more reliable term for channel is *inâonan.*

There are other variations of this name: *Wow-e-yat-ton-ong* (W. W. Warren), *Wauweeautonao* (Trowbridge), *Wawiiatanong* (Verwyst), and *Waweashton,* on a Canadian river (Lemoine). According to Verwyst, it meant "where the current of the river whirls around." Lemoine defined *Waweeashton* in Montagnais as "it is somewhat crooked."[7]

Ishkote Lake in Iron and Gogebic counties has the Ojibwa name for "fire." The reason for it is undetermined.[8]

Kalamazoo is the name of a river, city, and township in Kalamazoo County and of a lake in Allegan County. Kalamazoo County was established July 30, 1830, and was allegedly named from the "Indian" name of the river, *Ke-kanamazoo,* "the boiling pot." During the French period, this river was known by an older native name, *Meramec* (catfish), perhaps from the Potawatomi. There are several spellings and translations of Kalamazoo. In a treaty with the Potawatomi at Chicago, August 29, 1821, the river is called *Ke-kal-i-ma-zoo,* a word that may be from the Miami, because of the presence of *l.* "Boiling water" is the interpretation of Kalamazoo that is given on the roadside marker on I-94 south of the city of Kalamazoo, but others are found in the literature.

Father Verwyst called this name a "corruption of Kikanamoso (it smokes, or he is troubled with smoke—e.g., in his wigwam, pr. *kee-kah-nah-mo-zo* or *kee-kau-nau-mo-zo.*" Concurring, W. R. Gerard maintained that Kalamazoo is a slight alteration of old Ojibwa *kikalâmoza,* meaning "he is inconvenienced by smoke in his lodge." Schoolcraft's earlier view that this name was from *negikanamazoo,* referring to "otters beneath the surface," was branded by Gerard as an "etymological absurdity." It is the present writer's view that all previous theories of the name deserve this label. So also does the view that the name means "mirage or reflecting river."[9]

One of the following interpretations appears more credible. Kalamazoo could be a corruption of Ojibwa *kikikamagad,* "it goes or runs fast." It could also be a mangled form of *Kalimink,* a creek name in Ingham County, or *Killomick,* one of several early names of Indiana's Calumet River, each of them, perhaps, having lost terminal syllables. The latter name has been interpreted by Jacob P. Dunn as "deep, still water." Forms in *l* are from the Miami, the others from Potawatomi or Ojibwa. There are different characteristics in these two interpretations, "fast" and "still," but the river could fit each of them in different parts of its course.[10]

In Palms Book State Park, Schoolcraft County, Kitchi-ti-ki-pi Spring gushes forth, pouring sixteen thousand gallons of water per minute into Indian Lake. The official translation of this name, from Ojibwa, is "Big Spring." *Kitchi* means great, but *tikipi* appears to be a corruption from the Ojibwa terms in Baraga's dictionary, where *takig* signifies "spring" or "fountain," and *takigama* means "spring water."[11]

The Manistee and Little Manistee rivers empty into Manistee Lake at the city of Manistee in Manistee County. The name is also given to a township in that county, to a lake in Kalkaska County, and to a national forest that occupies parts of nine counties. Little Manistee is a village in Lake County. The earliest explanation of this name is that the Indians interpreted it as "River at whose mouth there are islands." However, there are no islands at the river's mouth today, nor are any shown on older maps. Other explanations in print are "island or island in the river" and "spirit of the woods." If this name is not related to Manistique, which we consider next, it might be translated as "small island in a river," from *minitig,* with a diminutive *s.*[12]

However, early maps indicate that the name Manistee is related to that of Manistique River, which joins Lake Michigan at Manistique, in Schoolcraft County. La Potherie, in his account of the "Adventures of Nicolas Perrot" (1665–70), called the present Manistique River *Oulamanistik.* The Manistee River, on the other side of Lake Michigan, is shown as *Aramoni* on Minet's *Carte de Louisiane* (1685), as *Manistigua* on Coronelli's map of 1688, *R. d'Oulamithe* on a map published by Charlevoix in 1744, and *Oulamaniti* in D'Anville's maps of 1746 and 1745.[13]

The *Oulaman* in some of these names was also attached, in variant spellings, to the Nipigon River (Ontario), which is considered below. In various spellings it appears in Algonquian placenames reaching from Quebec (Aramoni) and Maine (Olaman) to Minnesota (Onamia). All are variants of *onaman,* the Ojibwa word for "vermilion." The presence of *l, n,* or *r* in its various forms is accounted for sometimes by dialectic or language differences and sometimes by the hearing impressions of European recorders.

Part of Map of New France by Jacques Bellin, Royal French Marine Engineer, 1755. Lakes Michigan and Huron are shown with their present spelling. Perhaps the clearest and most detailed of early maps showing the region that became the state of Michigan, it has many errors. There are nonexistent islands in Lake Superior and an imaginary ridge in the Lower Peninsula.

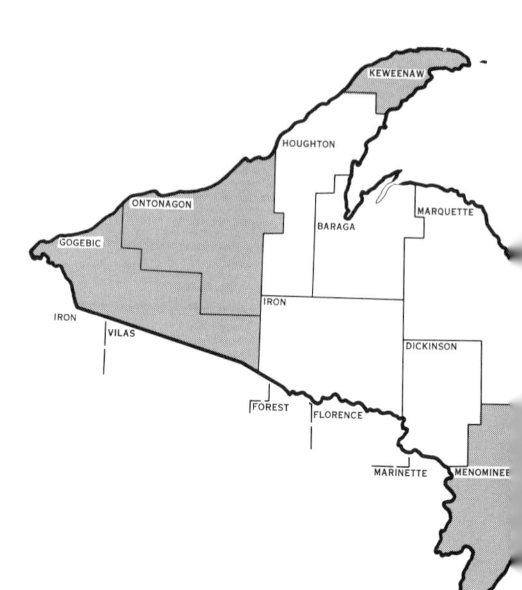


Aboriginal Checkerboard. Thirty-three of Michigan's eighty-three counties have Indian, semi-Indian, or pseudo-Indian names. (Base map © Rand McNally & Company. Used by permission.)

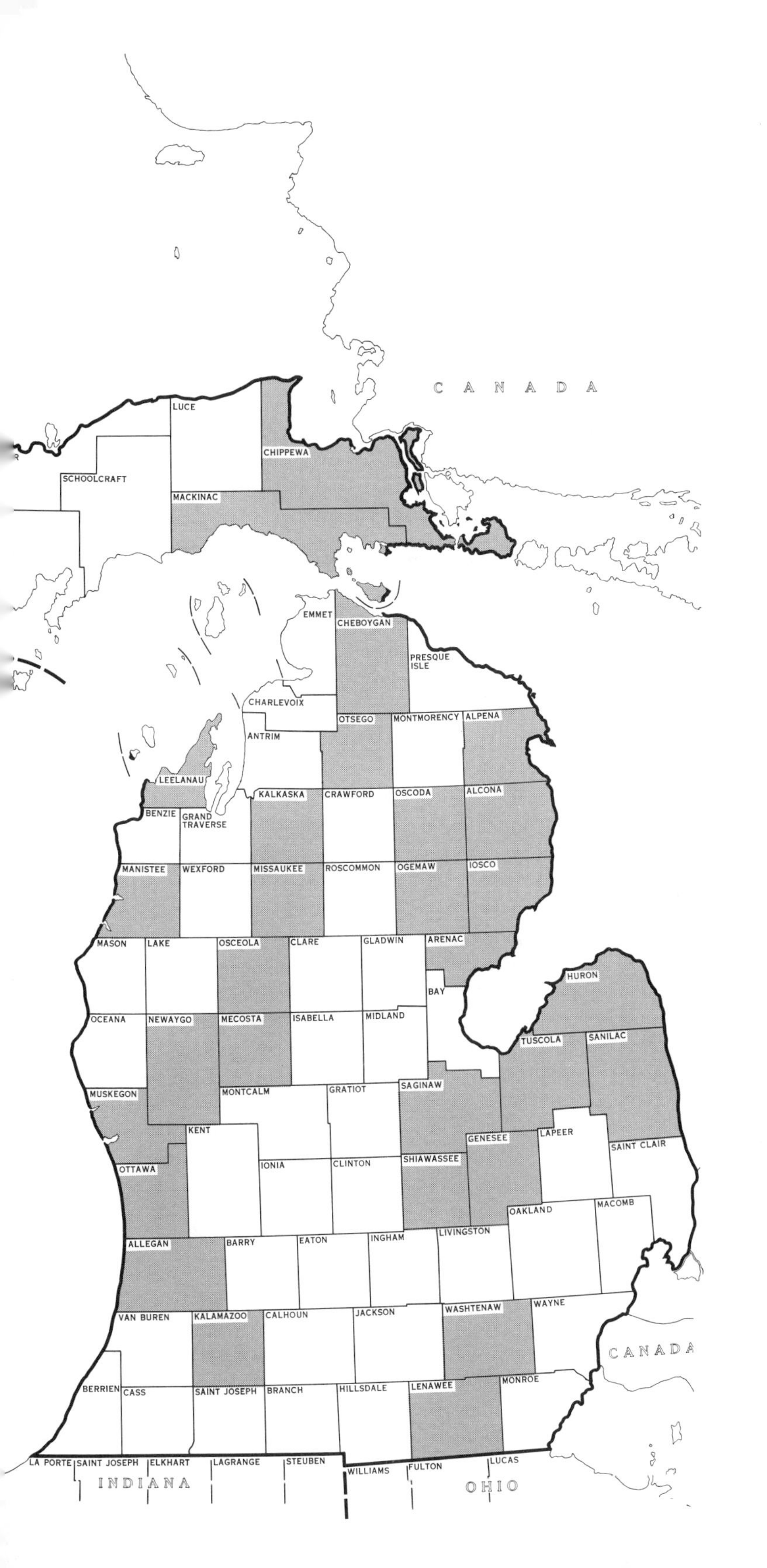

CANADA
LUCE
CHIPPEWA
SCHOOLCRAFT
MACKINAC
EMMET
CHEBOYGAN
PRESQUE ISLE
CHARLEVOIX
OTSEGO
MONTMORENCY
ALPENA
ANTRIM
LEELANAU
KALKASKA
CRAWFORD
OSCODA
ALCONA
BENZIE
GRAND TRAVERSE
MANISTEE
WEXFORD
MISSAUKEE
ROSCOMMON
OGEMAW
IOSCO
MASON
LAKE
OSCEOLA
CLARE
GLADWIN
ARENAC
HURON
BAY
OCEANA
NEWAYGO
MECOSTA
ISABELLA
MIDLAND
TUSCOLA
SANILAC
MUSKEGON
MONTCALM
GRATIOT
SAGINAW
KENT
GENESEE
LAPEER
SAINT CLAIR
OTTAWA
IONIA
CLINTON
SHIAWASSEE
OAKLAND
MACOMB
ALLEGAN
BARRY
EATON
INGHAM
LIVINGSTON
VAN BUREN
KALAMAZOO
CALHOUN
JACKSON
WASHTENAW
WAYNE
CANADA
BERRIEN
CASS
SAINT JOSEPH
BRANCH
HILLSDALE
LENAWEE
MONROE
LA PORTE
SAINT JOSEPH
ELKHART
LAGRANGE
STEUBEN
WILLIAMS
FULTON
LUCAS
INDIANA
OHIO

Thomas Jefferson's Plan for the Northwest Territory. The name Michigania would have been given to the region now occupied by Wisconsin and the Upper Peninsula, while the Lower Peninsula would have become the state of Cherronesus. (Courtesy The Library of Congress.)

Henry Rowe Schoolcraft (1793–1864), pioneer ethnologist, creator of many Michigan place-names. (Daguerreotype from the Charles F. Gunther collection; courtesy Chicago Historical Society.)

Topenibee (ca. 1740–1826 or 1840), of St. Joseph, principal chief of Michigan Potawatomi during most of his life. In 1881 his name was given to a village in Cheboygan County. (Courtesy Northern Indiana Historical Society.)

Chief Charles Kawbawgam (1799 or 1819?–1902), for whom Lake Kawbawgan (so spelled) in Marquette County is named. (From the collections of the Michigan State Archives, Department of State.)

Osceola (1803?–38), Seminole warrior, leader of the Florida Indians' resistance to removal, who died a prisoner. Literature about the Seminole wars gave him a heroic reputation, so that more places are named for him than for any other Indian. In Michigan his name is borne by a county, a village, and three townships. Osceola, from *assi-yahola,* means "black drink halloer," referring to the ceremonial beverage *cassine* or *yaupon,* and the cry accompanying its use. (From a painting by George Catlin, 1837; courtesy National Archives.)

Leopold Pokagon, "The Rib" (1775–1841), Potawatomi chief of St. Joseph. He is commemorated by the names of a village, township, creek, and prairie in Cass County. (Courtesy Northern Indiana Historical Society.)

Chief Okemos, "Little Chief" (1782?–1858), an Ojibwa Indian for whom the town of Okemos is named. He signed treaties in 1819 and 1833. (From the collections of the Michigan State Archives, Department of State.)

Supposed portrait of Tecumseh (1768–1813), Shawnee leader who died fighting Americans in the battle of the Thames, October 5, 1813. (From the collections of the Michigan State Archives, Department of State.)

The Indian custom of spearing fish at night by torchlight gave rise to several place-names: Torch Bay and Torch Lake, Houghton County; Torch Lake and Torch River in Antrim and Kalkaska counties; the towns of Torch Lake in Antrim County and Torch River in Kalkaska County; and (in French) Lac du Flambeau in Vilas County, Wisconsin. (From a painting by Paul Kane, "Spearing Salmon by Torchlight," courtesy Royal Ontario Museum, Toronto, Canada.)

Gagnieur (1918) gave this explanation of Manistique:

> *Onaman* (red ochre or vermillion) and *tigweia* (referring to a river), for the reason that along that river are to be found in plenty the "onaman" or vermillion roots. It is still called by the Indians *Onamanitikong,* i.e., Vermillion River.

Gagnieur did not identify the "vermillion roots." If named for roots, the reference could be to the bloodroot, *Sanguinaria candensis,* from which is named the Salamonie River of Indiana. More probably, the name refers to the availability of clay used in making vermilion paint. It is reported that English maps once called Manistique the Red Clay River.[14]

It appears likely that the name Aramoni, Oulamaniti, or Manistee River has the same origin as that of Manistique. One dissenting opinion explains Manistique as a derivation from Ojibwa *manistegweia,* "crooked river," while another calls it Algonquin for "island all wooded."[15]

The Manistique River has given its name to a town and township in Schoolcraft County, to Manistique and South Manistique lakes in Mackinac County, and to North Manistique Lake in Luce County.

Minnewanna Lake in Metamora State Park, Lapeer County, could receive its name from the Ojibwa term *miniwan,* "it produces or yields fruit." It could also be a perversion of *minnewawa,* a word from *Hiawatha* that appears on a waterfall in Ontonagon County. There is a more remote chance that it comes from the name of Minavavana, known to the French as Le Grand Saulteur ("The Big Jumper"), a chief who headed the Ojibwa of Mackinac during Pontiac's rebellion and released Alexander Henry to Wawatam.[16]

Nepessing Lake in Lapeer County has its name reversed in the name of Lake Nepessing, a village on its shores. This name is part of a considerable class of names containing the root word for water—*nipi,* and so on. Ontario has two lakes called Nipissing, while Nippesing was once (1878) the name of a hamlet in Isabella County, Michigan. Nippersink is the name of a creek and lake in McHenry County, Illinois, and related to these is the name of Neebish Island, of St. Marys River, considered elsewhere.

The first part of these names (*nepess, nipess, nippers*—the last very corrupted) simply means, in Ojibwa, "little water," as the root *nippi* is diminutized by the *s* ending, while *ing* or *ink* is a locative suffix, equivalent to "at" or "in."[17] The relationship to Ojibwa of other Algonquian languages is seen in the following:

> *nippissee* (dim. of *nippe,* small water), a pool or pond. Trumbull, *Natick Dictionary,* p. 85.
> *Nipissis* (Labrador place-name), *nipishish,* the small water. Lemoine, *Francais-Montagnais,* p. 280.

Point Nipigon on Lake Huron, Cheboygan County, is a name undoubtedly related to that of the town of Nipigon and Nipigon Bay, Lake, and River north of Lake Superior, in Ontario, and perhaps also Neepikon Falls, Gogebic County. The Ontario name is the residue of a once longer name. Father Claude Allouez, late-seventeenth-century Jesuit, called the lake Alimibegoung and said it was the place to which, according to Ojibwa mythology, the beaver fled from Lake Superior to escape from Michabou, the great hare. A century later, Jonathan Carver wrote of "the Nipigon River, or, as the French pronounce it, the Allanipegon, which leads to a band of the Chipeways, inhabiting a lake of the same name."[18]

The early writers did not translate the name, and their use of *l* seems to indicate a non-Ojibwa Algonquian origin of this name. The recent interpretations are "deep, clear water" and "continuous water" (from *animi-be-go-ong*). The name in its present form could be translated from Ojibwa as "at the water" or "place of water" (*nipi,* "water," + *gan,* "place") or "snow water" (*nipi,* "water" + *gon,* "snow").[19]

However, with the earlier prefixes a different original meaning is indicated. It is possible that these are mangled versions of Ojibwa *onaman,* "vermilion," and that the name signified "place of vermilion water." This view is somewhat reinforced by the fact that the prefixes *alimi, allini,* and so on, once joined to this name, resemble the initial syllables of the old name of Oulamanistik, now Manistique, River (q.v.), which Gagnieur translated as Vermilion River. Numerous "vermilion" place-names are found in other states, especially Minnesota. Vermilion paint was popular with the Indians, and clay for it could have been found in those locations, or the color of the water at sunset could have given rise to the name.[20]

Nichwagh Lake in Livingston County may have its name corrupted from Ojibwa *nijwak,* "two hundred." If that is so, the reason for the peculiar name is not known.[21]

The name of the village and township of Oshtemo in Kalamazoo County has been interpreted as "head waters." It is perhaps a mangled and shortened form of *oshtigwânima,* Ojibwa for "head," and apparently refers to a nearby headstream of the Kalamazoo River.[22]

Pemene Creek, a tributary of the Menominee River in Menominee County, and its waterfall, have a name that has been translated as "elbow," a name said to refer to a bend in the river. Pemene is a reasonable approximation of *biminik,* "elbow," in Baraga's Ojibwa dictionary. The Menominee term *pemin,* "sweat bath," is a closer approximation, but the name would be less fitting.[23]

Petobego Lake in Grand Traverse County, for which a state fish and wildlife area has been named, bears a name that is most closely related to Ojibwa *bitobig,* a pond or pool. The terminal *o* in the present name may be the remains of *ong,* a locative suffix.[24]

In our discussion of the state name, it was indicated that *sagaigan* was a generic Ojibwa term for an inland lake. On today's map, Sagaigan is the name of a creek and lake in Gogebic County.

The village and township of Sebewa in Ionia County take their name from Sebewa Creek, a tributary of the Grand River. This name is probably a mere variation of Shebeon, the name of a river that joins Saginaw Bay in Huron County. Both appear to come from *sibi-wan,* Ojibwa for "rivers, streams." There is no apparent evidence to support the alternative explanations, "running water" and "little creek."[25]

Closely allied to the preceding name is Sebewaing, the name of a town and township in Huron County. Eli Thomas (Little Elk), an elderly Ojibwa speaker at the Isabella Reservation, informed me on August 3, 1982, that this name means "where the river is." This closely conforms to two other explanations in print, but dismisses Romig's translation, "crooked creek."[26]

A verse of doggerel published in 1889 reads:

> To where the Saginaw's began
> The Shiawassee there doth run,
> Its Indian name doth indicate,
> The river that is running straight.

Shiawassee is the name of a river, tributary to the Saginaw, as well as of a city, county, township, state game area, and national wildlife refuge. From the river came also the name of a pond in

Shiawassee County, lakes in Genesee and Oakland counties, and the village of Shiawasseetown in Shiawassee County. Eli Thomas told me that this name was too corrupted to be translated. The first volume of the *Michigan Pioneer and Historical Society Collections* included the interpretation given in the above-quoted verse. However, Dwight Kelton (1888) gave the credible argument that it signifies "it runs back and forwards," or "the river that twists about." That remains the generally accepted interpretation. Another explanation, "green," involves too much corruption to be acceptable.[27]

Tahquamenon is one of the most difficult names in Michigan to explain. The "rushing Tahquamenaw" of *Hiawatha* is a river that discharges into Lake Superior in Chippewa County. It is noted for two spectacular waterfalls enclosed within Tahquamenon State Park. The name has also been attached to Tahquamenon Bay on Lake Superior.

Father Gagnieur asserted (1930) that the original form of this name was Otikwaminang but that his Indian informants could not explain it, adding that "it must be therefore a very old word, and the original meaning appears to be in the long past."[28]

The official brochure for Tahquamenon State Park translates the name as "Marsh of the blueberries." *Menon* indeed could be from *menan,* the Menominee word for blueberries, but *tahqua* does not mean marsh in any language of this region, and blueberries do not grow in marshes. The name as it now stands could be connected with *takwa,* Ojibwa for "short," and *manan,* defined by Baraga as "cornet tree," *bois dur.* However, such a name makes little sense for a river, and such piecemeal analysis is hazardous.

Others hold that the name refers to "dark waters" and "black-gold waters colored by miles of peat swamps." These views are not sustained by vocabularies of the region.[29] DeLisle's map of 1703 shows this name as Outakwamenon, and also places Villages d'Outaouacs (Ottawas) in that vicinity.[30] Possibly *outakwa* in the place-name is a misspelling of *outaouac,* and *menon* means "good land or place," as in Minong (q.v.). All of these suggestions are conjectural, and Gagnieur may be correct in concluding that the original meaning of this name is lost.

The Tittabawassee River joins the Shiawassee to form the Saginaw River at the city of Saginaw. A township of Saginaw County is named for it. Eli Thomas told me that this name means

"The river runs along the line." Albert Miller's verse of 1889 gave essentially the same explanation:

> The Indian name is pronounced, I think
> As if it were written Ta-ta-ba-wa-sink,
> Which is the same as if you say,
> Running parallel with the bay;
> That you will see is the river's course,
> Which gives the Indian name its force.[31]

On the meaning of this name, all writers are in substantial agreement. Ephraim Williams wrote (1886) that whites at first called this stream the Tiffin River but "changed it to the Indian name, Ta-tu-ba-war-say, the meaning of which is the river running around the shore, as it does make off around Saginaw Bay and Lake Huron."[32] James Lanman translated the name (1839) as "river that runs alongside," while Fred Dustin (1928) rendered it "river that follows the shore." Among variant spellings of this name are *Tet-abawasin*, in the Treaty of Saginaw, September 24, 1819, and *Te-ti-pe-wa-say* in Blackbird's *History*.[33]

The Two Hearted River and the Little Two Hearted River, tributaries of Lake Superior in Luce County, were reportedly so named by traders from their Indian name. It is said that the Ojibwa name was "*Nizhode* (two or twin) *sibi*, so named because their outlets were near together." The Ojibwa name for "heart" is *odeima* and does not appear in this name, so it may be a white man's interpretation. However, Haines said that the original form of the name was *Neezhoda seepee*, which was broken down into *neezh* (*nij* in Baraga), "two," + *oda*, "heart," and *seepee*, "river." However, Thomas McKenney wrote in his journal for July 12, 1826, that he "entered *Twin*, or as it has been, though not so well, called 'Double hearted' river." S. Mitchell's map of 1839 labeled the main stream "Twin River."[34]

## Notes

1. Dustin, "Some Indian Place-names," 733; Miller, "Rivers of the Saginaw Valley," 504; Schoolcraft, *Personal Memoirs*, 362.
2. Fox, "Place Names of Berrien County," 13; Haines, *American Indian*, 722; Orville W. Coolidge, *A Twentieth Century History of Berrien County, Michigan* (Chicago: Lewis Publishing Co., 1906), 276.

3. Fox, "Place Names of Berrien County," 32; idem, "Place Names of Cass County," 467; Gannett, *American Names,* 108; Romig, *Michigan Place Names,* 162; Clifton, *Prairie People,* 233; Gailland, "English-Potawatomi," 25.
4. Verwyst, "Geographical Names," 391; Romig, *Michigan Place Names,* 227, 313; Haines, *American Indian,* 729; Baraga, *Otchipwe Language* 1:214.
5. Baraga, *Otchipwe Language* 1:120; "Reports of Counties," 326; R. V. Williams, "Washtenaw County," *MPHSC* 4 (1881): 393; Schoolcraft, *Personal Memoirs,* 595; Goss, "Indians of the Grand River Valley," 182.
6. Verwyst, "Geographical Names," 398.
7. Baraga, *Otchipwe Language* 1:47; 2:406; Schoolcraft, *Indian Tribes* 5:594; Verwyst, "Chippewa Names," 259; Lemoine, *Dictionnaire,* 281.
8. Baraga, *Otchipwe Language* 2:158.
9. "Reports of Counties," 200; Kappler, *Indian Treaties,* 201; Verwyst, "Geographical Names," 391; William R. Gerard, in "Anthropological Miscellany," *American Anthropologist,* n.s., 13 (April–June, 1911): 337; Romig, *Michigan Place Names,* 297.
10. Baraga, *Otchipwe Language* 2:186; Dunn, *True Indian Stories,* 250.
11. Official brochure, Palms Book State Park, Michigan Department of Conservation; Baraga, *Otchipwe Language* 1:242.
12. "Reports of Counties," 259; Baraga, *Otchipwe Language* 2:242.
13. Kellogg, *Early Narratives,* 79; Tucker, *Indian Villages,* plates 7, 10; Karpinski, *Historical Atlas.*
14. Baraga, *Otchipwe Language* 1:278; Gagnieur, "Indian Place Names" (1918), 550–51; Jenks, "County Names of Michigan," 468.
15. Verwyst, "Geographical Names," 392; Assiniwi, *Lexique des Noms Indiens,* 66.
16. Baraga, *Otchipwe Language* 2:242; Greenman, "Indian Chiefs," 220, 227–28.
17. Romig, *Michigan Place Names,* 399; Vogel, *Indian Place Names,* 89; Baraga, *Otchipwe Language* 1:300.
18. Kellogg, *Early Narratives,* 144; Carver, *Travels,* 137.
19. J. A. Rayburn, "Geographical Names of Amerindian Origin in Canada, Part II," *Names* 17 (June, 1969): 151; W. B. Hamilton, *Macmillan Book of Canadian Place Names* (Toronto: Macmillan of Canada, 1978), 191; Baraga, *Otchipwe Language* 1:194; 2:141.
20. Baraga, *Otchipwe Language* 1:278; Gagnieur, "Indian Place Names" (1918), 550–51.
21. Baraga, *Otchipwe Language* 2:290.
22. Romig, *Michigan Place Names,* 421; Baraga, *Otchipwe Language* 1:128.
23. Kuhm, "Indian Place-Names," 101; Baraga, *Otchipwe Language* 1:85; Hoffman, "Menomini Indians," 310.
24. Baraga, *Otchipwe Language* 1:196.
25. Ibid., 2:366; Haines, *American Indian,* 780; Romig, *Michigan Place Names,* 504.
26. Verwyst, "Chippewa Names," 260; FWP, *Michigan,* 455; Romig, *Michigan Place Names,* 505.
27. Miller, "Rivers of the Saginaw Valley," 504; Kelton, *Indian Names of Places,* 50; Walton, "Indian Place Names," 26; Romig, *Michigan Place Names,* 512;

Kane, *American Counties,* 333; Dustin, "Some Indian Place-names," 732. It is reported that a local Ottawa chief named Shiawassee lived north of Grand Haven some time prior to the Civil War. Angered by land cessions, he migrated to Canada. There is no evidence linking him to the Shiawassee place-names. William N. Ferry, "Ottawa's Old Settlers," *MPHSC* 30 (1905): 577–78.

28. William F. Gagnieur, "Tahquamenon," *Michigan History* 14 (July, 1930): 557.
29. Hoffman, "Menomini Indians," 316; Baraga, *Otchipwe Language* 1:216, 228; Romig, *Michigan Place Names,* 548; Hamilton, "Chippewa County Place Names," 642.
30. Karpinski, *Historical Atlas,* passim.
31. Miller, "Rivers of the Saginaw Valley," 501.
32. Ephraim Williams, "Remembrances of Early Days," *MPHSC* 10 (1886): 142.
33. James H. Lanman, *History of Michigan* (New York: E. French, 1839), 266; Dustin, "Some Indian Place-names," 731; Kappler, *Indian Treaties,* 185; Blackbird, *History of the Ottawa and Chippewa,* 60.
34. Walton, "Indian Place Names," 32; Haines, *American Indian,* 759; Baraga, *Otchipwe Language* 1:131, 272; 2:289; McKenney, *Sketches,* 178.

# XV

# Topography and Geology

The most prominent feature of the south shore of Lake Superior is the peninsula of Keweenaw. Its curved claw juts seventy-five miles into the lake, from southwest to northeast. Before the time of aids to navigation, it was a graveyard of ships.

The tip of the peninsula is called Point Keweenaw. In order to avoid its hazards and to cut the distance, canoes and small boats left the lake at a point some fifteen miles above L'Anse (if upbound), following the Portage River and Portage Lake to the site of present Houghton, where a short portage was made to a stream leading to the other side.

Thomas L. McKenney was traveling westward by barge on July 19, 1826, when he reached "Point Kewewana" and wrote in his journal:

> This point is forty-five miles [*sic*] from what is called the Once [L'Anse], a bay which washes its shores where it is united with the mainland. On doubling it, we shall have to coast the same distance on its eastern side, to the *Portage,* the carrying place for all the trade of this place, except such as goes up in barges. The Indians, and those who go in canoes, never come round this point. Our barges must do so, and we have coasted it to keep company, and to get information. Our Indians continued along the shore opposite the Huron Islands, and crossed the portage. We shall cross this carrying place on our return, as the Governor means to return in a canoe.[1]

McKenney made a distinction, as the earliest maps do, between Keweenaw Point and the portage some sixty miles below it. The distinction is important, because many writers have translated Keweenaw as "portage." The Ojibwa name for a portage, however, is *onigam.* Its ancient application to the portage route across the peninsula is recalled by the name of Onigaming ("Portage Place")

Supper Club on Portage Lake near Houghton. Onigum is the name of a Chippewa village at Leech Lake, Minnesota, which also occupies a portage site.

Father Verwyst (1892) connected Keweenaw to *kakiweonan,* "I cross a point of land by boat," and gave the same explanation for the name of Kewaunee, Wisconsin. Father Baraga gave the same form, *kakiweonan,* but translated it as "The place where they traverse a point of land on foot." Neither writer explained their dropping the initial syllable *ka.* There is some similarity between Keweenaw and Baraga's phrase *nin giwewina,* "I carry (lead or convey) him (her, it) back again." Several later writers have adopted the explanations of Baraga or Verwyst, although they are subject to serious objections, aside from the question of the missing *ka.*[2]

The most serious objection is that early maps distinguish between the tip of the peninsula and the portage route. The Jesuit map of 1671 shows *Pointe de Kiaonan* at the end of the peninsula, *Kiouehounaning* at a nonexistent inlet about halfway toward the Portage River, *Portage* at the west end of the Portage River, and *Ance de Kiaonan* at present L'Anse, at the bottom of the peninsula. Franquelin (1688) shows *Pointe Kiouenan* at the end of the peninsula and *Portage* at the Portage River, about sixty miles via coastline to the southwest. Again, Jacques Bellin's map (1755) shows *Pointe de Kioueounan* at the northeast extremity of the peninsula and Portage near its base. The M. Lewis map (1806) shows *Keonan* Point only. The first source to identify the name Keweenaw with the Portage River was Jedidiah Morse (1822), who labeled the stream *Quawionone,* or *Keweena* River. Until that time, no one had given a translation of the name. The first to do so was Joseph N. Nicollet, who in his journal of 1836 mentioned "Kiwewina (the bend)." No other explanation appeared in print for forty years.[3]

Nicollet's translation was probably obtained from the Indians, and since it is the earliest one it deserves great weight. Moreover, this explanation fits the feature to which it is applied on the early maps, for the peninsula is indeed a "bend," and this name was not applied to the portage site until 1822. Finally, Nicollet's other name translations have usually proven to be correct.

An alternate translation given by Haines (1888) is close to Nicollet's: "from Kewaywenon, signifying a detour or returning around a point." Baraga's word for detour is *giwedeonan.* If *k* is substituted for *g,* the term becomes very similar to that of Haines.

Baraga tends to use *g* for *k*, as when he renders *kewadin* (north) as *giwedin.* According to Verwyst, the Ojibwa name for the site of present De Tour Village, Mackinac County, was *Giwideonaning,* i.e., "point which we go around in a canoe." Going around is much different from crossing or portaging and could be an acceptable description of the circumstances, if not the name, of Keweenaw. Either "the bend," as given by Nicollet, or "detour," as suggested by Haines, seems to be the best explanation of the name Keweenaw.[4]

The name of Keweenaw was given to the bay on the east side of the peninsula, to the Indian village of Keweenaw Bay on the bay's western shore, and to the county at the northeast end of the peninsula, established March 11, 1861.

Minong was the aboriginal name of Isle Royale in Lake Superior and was used on European maps in the late seventeenth century and much of the eighteenth. However, the name Isle Royale appears as early as 1744 on a map by Bellin. Today the name Minong survives on a ten-acre island off the shore of Isle Royale and on Minong Ridge, which forms the backbone of the island. Formerly a copper mine, resort, and post office on the island bore this name. A lake, village, and township in Washburn County, Wisconsin, are also named Minong. With reference to Isle Royale, Verwyst wrote that "The Chippewa name for this island is pr. *mee-nong,* and means a good, high place." The same name in Wisconsin, according to some, could signify "blueberry place."[5]

In Lake Superior immediately offshore from Munising in Alger County is a large wooded island called Grand Island. The Indians called this place *Gitchi* (or *Kitchi*) *Minissing,* meaning "at the big island." From that came the name of the town, township, bay, and waterfall called Munising, a name only slightly altered from the original Ojibwa.[6]

Joseph Nicollet wrote in his *Report* (1841) that "there is what the Chippeways call a *mashkeg,* and the Sioux *wiwi,* or swamp, or elastic prairie." From this term, also written as *maskig* and *muskeg,* came a tribal name: Muskegoes (Swampy Crees). Moreover, the word *muskeg* was adopted into English as a term for spongy, waterladen ground and was used in numerous place-names here and in Canada. Muskeg Lake in Gogebic County is one of its many applications. A place called Maskigo is mentioned in a treaty with the

Ottawa, March 26, 1836. Today it is called Muskegon, from Ojibwa *maskig,* "swamp or marsh," and *ong,* "place."[7]

The name of the place comes from the Muskegon River, which rises in the marshes of Missaukee and Roscommon counties and discharges into Muskegon Lake at Muskegon, and from thence into Lake Michigan. Along the way, in Newaygo County, it receives the Little Muskegon River. Besides the city of Muskegon, the name of the river has been given to Muskegon Township and County, to the suburbs of Muskegon Heights and North Muskegon, and to Muskegon State Park.

Nayanquing Point on Saginaw Bay, Bay County, is probably named for its position midway between Saganing and the mouth of Saginaw River. Its root appears to be from Ojibwa *nawaiiwan,* "it is the middle, the centre."[8]

Neebish is the name of the second largest island in St. Marys Channel, Chippewa County. When Father William Gagnieur wrestled with this name in several articles, he came up with a bewildering variety of interpretations: bad water, dirty water, tea, leaf, and foliage. He added yet another view, saying that the Ottawa called an inland lake *nibish.* Finally, Elijah Haines chose "bad rapids" as his interpretation, while Romig offered "where the water boils."

Part of the problem arises from alleged older spellings of the name—e.g., *anibish.* If we analyze the present name, it appears as simply an aberrant spelling of *nibish* or *nebeesh,* a word for "water" in Ojibwa, Ottawa, and Potawatomi. Often *nibi* or *nipi* is found as the root of water names, such as Nipigon and Nipissing (q.v.). There are also Nebish lakes in Vilas County, Wisconsin, and Beltrami County, Minnesota.[9]

The village of Onekama in Manistee County has its name explained in three ways. One of them can be rejected at once: "place of contentment." The second, "portage," is probably given because of its similarity to *onigam,* Ojibwa for "portage," and its location on Portage Lake. The third view, by Verwyst, connects the name with Ojibwa *onikama,* "an arm." The name is appropriate, for Portage Lake is an arm of Lake Michigan.[10]

Pequaming is a little village on the north edge of L'Anse Indian reservation, Baraga County. It is situated on a peninsula called Pequaming, jutting into Pequaming Bay, an arm of Keweenaw Bay. Two published explanations of this name are "Point Village or Cape Point" (from Pequaquawaming) and "wooded peninsula." However, the name has nothing to do with a point of land (*neiashi*) or a peninsula (*gigawekamiga*). The name is quite obviously derived from *bâgwa* (or *pagwa*), meaning "shallow," plus *ing*, "at"; thus, "at the shallow place." The name is akin to those of Pakwash Lake, Ontario, translated as "shallow," and Pugwash, Nova Scotia, meaning the same thing, from Micmac *pagwesk.*[11]

In northeastern North America in Indian and pioneer days, a major means of travel was by way of Indian canoe routes. Wherever a river flowing in one direction came near another going the desired way, canoes and baggage were carried across the narrow divide between them, and the journey was continued. The place and the process were called by the French name *portage*, as described by William Keating in 1823:

> As soon as a canoe reaches a portage, a scene of bustle and activity takes place, which none can picture to themselves but such as have seen it. The goods are unloaded, and conveyed across, while the canoe is carried by the stern and bowsman. As soon as they have reached the end of the portage, it is launched and reloaded without any loss of time.

A canoe party could travel from the foot of Lake Erie through Lake Huron and Michigan, and thence to the Mississippi and south to the Gulf of Mexico by making a short portage at present Portage, Wisconsin, or Chicago, Illinois. Both of these portages and others were used by Marquette and Joliet in 1673. From Montreal, however, canoe parties traveled to Lake Huron by way of the multiportage Ottawa River, in order to avoid the Iroquois.

Portages were also used to go around rapids or waterfalls or to shorten a coastal route by crossing a peninsula, as at Keweenaw, previously mentioned. The English at first called these portages "carrying places," but eventually the French name replaced theirs. The old portage routes were once considered important strategic points and were frequently mentioned in Indian treaties. The portage trails eventually became the routes of highways and canals. At portages forts and trading posts were built that grew into cities such as Grand Rapids, Houghton, and Sault Ste. Marie, Michigan, South Bend, Indiana, and Chicago, Illinois. The former importance of portages is shown by the large number of Portage place-names. The United States Geological Survey lists twenty-nine in Michigan alone.

As noted earlier, the portage route across the Keweenaw Peninsula is today an all-water passage by canal. The waters connecting the waters on two sides of it are today called Portage Lake and the Portage River. Portage Bay, in Delta County, marks the portage across the Garden Peninsula from Lake Michigan to Big Bay de Noc. Portage Township in Mackinac County is named for the portage between North and South Manistique lakes. Portage Creek, Portage Lake, and Portage Point mark a canoe route in Manistee County. Portage Lake in Washtenaw and Livingston counties is on the Huron River canoe route, while Portage Lake, Little Portage Lake, and the Portage River in Jackson County mark a canoe route connecting with the Grand River. The Portage River rises in Kalamazoo County, passes through Portage Lake in St. Joseph County (where we have also the village of Portage Lake) and unites with the St. Joseph River at Three Rivers. Upriver at Mendon the St. Joseph also receives Little Portage Creek. Shallow water in these small streams necessitated frequent portages in late summer. The city of Portage in Kalamazoo County marks a canoe passage to the Kalamazoo River.

The Indian names for these portages have not survived as place-names, but all of them are a record of the Indian imprint on American travel history, and the nomenclature of these places is really theirs in translation.[12]

Potaganissing Bay is at the west end of Drummond Island in Lake Huron. Father Gagnieur called this name an archaic Indian word signifying "gaps." Charlotte Hamilton, concurring, said that "The Indians name for the island and adjacent bay was Potagannissing, referring to the numerous gaps in the archipelago."[13]

The Saganing River flows into Saginaw Bay at the Ojibwa community of Saganing, Arenac County. The name refers to "an outlet, the mouth of a stream." Baraga offered *sâging,* "at the mouth of a river." It is apparently a variation of Saginaw.[14]

The Saginaw River falls into Saginaw Bay, and from them come the names of the city and county of Saginaw. The meaning of this name is hotly debated. One widely held view is that Saginaw River and Bay are named for the Sauk Indians, who are said to have occupied the area before the arrival of the whites. According to

Blackbird (1887), "Saginaw is derived from the name [of the] O-saw-gees, who formerly lived there." The other main view is that Saginaw means "outlet of a river," and some maintain that the Sauks were named Saukie-uck, "people of the outlet," because of their residence there, instead of the place being named for them. This view originates with Cadillac (1718), who wrote that "The Sauk tribe is so called because Sauky means mouth of the river."[15]

However, the correct name of the Sauks is Osaukie-uck, "people of the yellow earth," according to their own traditions. The Indian agent to the Sauks, Thomas Forsyth, informs us (1827) that "The original and present name of the Sauk Indians proceeds from the compound word Sakie, alias, A-saw-we-kee literally Yellow Earth." The older spellings of the tribal name are preserved in the names of Osakis, Minnesota, and Ozaukee County, Wisconsin.

In 1892 Father Verwyst called Saginaw a "corruption of Os-aginang (place where the Sacs used to live)," but later (1916) stated merely that the name came from Ojibwa *saging* or *saginang*, "at the mouth of a river."[16] His later view is probably the correct one. It is notable that none of the many recorded variant spellings of Saginaw include the initial sound of *o* or *a*, although that was a part of the Sauk tribal name. If the Sauk lived about Saginaw Bay, as tradition says they did, the similarity between the tribal name and the place-name appears to be coincidental. Saginaw Bay was named, according to J. H. Trumbull, "from the mouth of the river which flows through it to the lake." The name appears to be closely related to several others: Saugago Lake, in Kent County (perhaps from *sâgaki,* "it comes out of the ground"), Saguenay in Canada, and Saugatuck in Michigan. The latter name, however, is imported from Connecticut.[17]

The name Saginaw occurs again at Saginaw Point and Mt. Saginaw on Isle Royale, but there it appears to be merely a borrowed name. Saginaw is also a transfer name in eight states.

The Ojibwa word for a bay was *wikwed.* Their name for the bay at L'Anse was, as Warren recorded it, *We-qua-dong.* The *ong* in this name was a prepositional suffix. The aboriginal name did not stick to that place, as the whites took the name Keweenaw from the tip of the nearby peninsula and gave it to the bay. However, the Ojibwa name for the bay was not confined to that place; it was a generic name, applied also to Grand Traverse Bay in the Lower

Peninsula. They diminutized it to *Wikwedonsing* (at the little bay) and gave it to Little Traverse Bay in Emmet County.

In 1886, at a resort just east of Harbor Springs, a post office was established with the aboriginal name of the bay, slightly altered to Wequetonsing. The post office was closed in 1957, but the name of the community continues, even though it is omitted from some maps. It can be seen on a golf course in that community.[18]

## Rocks and Minerals

Michigan's Upper Peninsula has long been known for its copper and iron deposits. The first, worked by Indians from ancient times, are now practically gone, while the second is much depleted. Indians did not use the iron, but they had words for both of the metals. Each of them survives in current place-names due to adoption by whites. Bewabic State Park in Iron County has a name faithfully approximating Baraga's term for "iron," *biwabic.* A variant spelling of this name is Pewabic, which was once the name of a mining settlement in Dickinson County and of a post office and township in Ontonagon County. The unincorporated community of Pewabic remains in Houghton County, while Pewabeck Falls, in the Little Iron River, has the name in variant spelling.

The presence of iron and copper in the Upper Peninsula was known to the earliest French explorers, for the Iron River in Ontonagon County was shown on the "Jesuits map" of the 1660s as "R. Du Piouabic ou du cuivre" ("Iron or Copper River"). *Piouabic* was their orthography for Ojibwa *pewabic* (iron) while *cuivre* was French for "copper."

*Miskwabic,* "red metal," was the Ojibwa name for "copper." It is recognizable in the name of Miscowaubic Peak in Ontonagon County. Also in Ontonagon County, the name of Mishwabic State Forest is undoubtedly meant to be Miskwabic.[19]

Wabeek Lake in Oakland County, as noticed earlier, seems to have taken its name from *wawbeek,* "a rock," as given in *Hiawatha* by Longfellow.

A rocky place was called *ajibikoka* by the Ojibwa. The word appears slightly altered in the name of Ajibikika Falls in Sucker Creek of the South Branch of the Ontonagon River, Gogebic County. *Kakabika* is an Ojibwa generic term for a steep, rocky waterfall. It is seen in the name of Kakabika Falls in Thayer Creek, a tributary of the South Branch of the Ontonagon River, Gogebic County. In almost identical orthography is the name of Kakabeka Falls, also named by the Ojibwa, in the Kaministiquia River near Thunder Bay, Ontario.[20]

Like the whites, the Ojibwa differentiate rock (*ajibik*) from stone (*assin*). At times the distinction is lost in translation, as in this old verse:

> From Bad river's mouth twelve miles will bring
> Us to the place Che-as-sin-ing;
> In the language of the Chippewa race,
> Its meaning is the Big Rock's place.

In court testimony at Saginaw in 1860, Chief Okemos spoke of "the Big Rock on the Shiawassee, called Chesaning." It was a large limestone rock in the river at the site of the present town of Chesaning, in Chesaning Township, Saginaw County, which is named for it. An Ojibwa village once stood there, on a tract called Big Rock Reserve. According to Eli Thomas (Little Elk) of the Isabella Reservation, who was interviewed by the author on August 3, 1982, the name Chesaning is from *kitci-ᵃsaning,* meaning "big rock," and the rock referred to was an object of veneration to the Indians.

It is reported that early settlers, with typical indifference to Indian sensibility, blasted the rock and burned it for lime. Only its name remains, more durable than the rock itself, furnishing an illustration in support of Reginald Bolton's remark: "These Indian names often tell us something of the features of a place which may have disappeared long ago."[21]

In 1839, it is said, two large stones stood near the mouth of the Devil's River, at the site of present Ossinike, adjacent to Lake Huron's shore in Alpena County. According to legend, Chief Shingabaw had told his people that after his death his spirit would return to this place, called *Wawsinika,* or "Image Stones." Hostile tribesmen, the story goes, loaded the stones into a canoe along with several captives and started across Thunder Bay. The offended spirits caused the water to boil, and the raiders were drowned. When the captives returned to shore, they found the stones restored to their former places. It is told that they were later appropriated by a fisherman as anchors for his nets and were lost in Lake Huron.

In 1848, a village was platted at the site of the image stones, and its name, Ossinike, was adapted from the Ojibwa name of the stones. The name was extended in time to the township and to a county park.

Slightly different interpretations of the name are on record. Verwyst called it a "corruption of *assinika,* (he gathers stones)." Haines called *Ossineke* "stony land," which closely matches the form *assineka,* defined by Baraga as "there are stones." *Assin* is "stone," and *aki* is "land." Related names outside Michigan include Assinica Lake in Quebec (probably from Montagnais), Assinika Lake in Cook County, Minnesota (Ojibwa), and Assinika Lake in Manitoba (Cree).[22]

"The Red Man," wrote one scholar, "gave names, which meant something, to the mountains, rivers and creeks."[23]

## Notes

1. McKenney, *Sketches,* 201.
2. Verwyst, "Geographical Names," 392; Baraga, *Otchipwe Language* 1:43; Romig, *Michigan Place Names,* 303; "Reports of Counties," 218.
3. Tucker, *Indian Villages,* plates 1, 31; Karpinski, *Historical Atlas;* Morse, *Report,* appendix, p. 28; Joseph N. Nicollet, *The Journals of Joseph N. Nicollet* (reprint, St. Paul: Minnesota Historical Society, 1970), 153.
4. Haines, *American Indian,* 738; Baraga, *Otchipwe Language* 1:72; Gagnieur, "Indian Place Names" (1918), 533.
5. Baraga, *Otchipwe Language* 2:244; Romig, *Michigan Place Names,* 288, 373; Dustin, "Isle Royale Place Names," 709; Verwyst, "Geographical Names," 393; Kuhm, "Indian Place-Names," 68.
6. Baraga, *Otchipwe Language* 1:119; FWP, *Michigan,* 562; Romig, *Michigan Place Names,* 385.
7. Joseph N. Nicollet, *Report Intended to Illustrate a Map of the Hydrographic Basin of the Upper Mississippi River,* 26th Cong., 2d Sess., 1841, S. Doc. 237 (Washington: Blair & Rives, 1843), 14; Kappler, *Indian Treaties,* 454–55; Baraga, *Otchipwe Language* 1:167.
8. Baraga, *Otchipwe Language* 2:279.
9. Gagnieur, "Indian Place Names" (1918), 534–35; idem, "Indian Place Names" (1919), 418–19; idem, "Indian Place Names," *Michigan History* 9 (January, 1925): 110; Verwyst, "Geographical Names," 394; Haines, *American Indian,* 758; Romig, *Michigan Place Names,* 390; see also Hamilton,

"Chippewa County Place Names," 639; Baraga, *Otchipwe Language* 1:283; Pokagon, *Ogimawkwe Mitigwaki,* 103; International Colportage Mission, *A . . . Dictionary of the Ojibwa and English Languages* (Toronto: 1912), 109.

10. Romig, *Michigan Place Names,* 416; Verwyst, "Geographical Names," 398.
11. Romig, *Michigan Place Names,* 436; FWP, *Michigan,* 596; Baraga, *Otchipwe Language* 1:226; Rayburn, "Geographical Names," 2:151; Silas T. Rand, *Rand's Micmac Dictionary* (Charlottetown, Prince Edward Island: Patriot Publishing Co., 1902), 372.
12. Quote from Keating, *Narrative of an Expedition* 2:85; Blanche M. Haines, "French and Indian Footprints at Three Rivers on the St. Joseph," *MPHSC* 38 (1912): 386–97. The importance of the old canoe trails is told in Archer B. Hulbert, *Portage Paths: The Keys to the Continent,* vol. 7 of *Historic Highways of America* (Cleveland: Arthur H. Clark, 1903), and David Lavender, *Winner Take All: The Trans-Canada Canoe Trail* (New York: McGraw Hill, 1977).
13. Gagnieur, "Indian Place Names" (1918), 534; Hamilton, "Chippewa County Place Names," p. 640.
14. Baraga, *Otchipwe Language* 2:360.
15. Blackbird, *History of the Ottawa and Chippewa,* 94; Quaife, *Western Country,* 65.
16. Blair, *Indian Tribes* 2:183; Verwyst, "Geographical Names," 397; idem, "Chippewa Names," 270; the last view is supported in Baraga, *Otchipwe Language* 1:175.
17. Baraga, *Otchipwe Language* 1:301; 2:359; J. H. Trumbull, "The Composition of Indian Geographical Names, Illustrated from the Algonkin Languages," *Collections Connecticut Historical Society* 2 (1870): 31.
18. Warren, *Ojibway Nation,* 86; Baraga, *Otchipwe Language* 1:25; Verwyst, "Geographical Names," 398; Romig, *Michigan Place Names,* 590.
19. Baraga, *Otchipwe Language* 1:59, 146.
20. Ibid. 2:22, 179.
21. Ibid. 1:246; Miller, "Rivers of the Saginaw Valley," 505; Webber, "Indian Cession of 1819," 524; Dustin, "Some Indian Place-names," 735; Reginald P. Bolton, *Indian Life of Long Ago in the City of New York* (New York: Crown Publishers, 1972), 48.
22. FWP, *Michigan,* 485–86; Romig, *Michigan Place Names,* 421; Verwyst, "Geographical Names," 295; Baraga, *Otchipwe Language* 1:154, 246; Geographic Board of Canada, *Place-Names of Manitoba* (Ottawa: Department of Interior, 1933), 12.
23. Donehoo, *Indian Villages,* v.

# XVI
# Descriptive Names

The names of Indian villages, when they possessed any, were often descriptive of their locations. Some of these names were taken over by whites, who also gave descriptive Indian or European names to locations, whether Indians used them or not (for example, Ishpeming). Since Indian villages were few, however, most descriptive names were given to topographical features, and such names form the vast majority of their geographical names. Most of these are discussed in their appropriate place in this book. Some others that had multiple applications or did not readily fit into other categories are examined here.

The largest group of descriptive names refer to some physiographical characteristic of a place, and so each of them normally contains a noun and an adjective or adverb. Sometimes they also contain verbs, and when translated into English often form short sentences.

One example of a sentence-name is Abitosse Creek, in Gogebic County, which has an Ojibwa name meaning "it comes or arrives to the middle."[1]

The city and county of Escanaba in Delta County received their name from the Escanaba River. Reportedly it was once called Esconawba, while Skonawba is the spelling used in the treaty of March 28, 1836. Verwyst thought the name might be from *misconabe* (pronounced *mis-co-nau-bai*), which he translated "red man." There are several other interpretations, but the one most commonly given is "flat rock." This is not sustained by Ojibwa vocabularies, if we suppose that the name is from their language. In Ojibwa, "flat, or it is flat" is *nabaga;* while "rock" is *ajibik.* Baraga gives *tessâbik* and *nabagâbik* as terms for "Flat stone." Nevertheless, Bela Hubbard's map of 1840 shows "*Osqua na be konk sepe* (Flat rock R.)" emptying into Lake Superior in present Alger County. That is the present Rock River. The similarity of the first part of the name to Escanaba is striking.

There is another possible interpretation of Escanaba: that the name is from Ojibwa *eshkam nibiwa,* "more and more, increasing." A similar name, Ashkum, in Illinois, commemorates a Potawatomi chief whose name meant "more and more." The Treaty of Washington, March 28, 1836, also lists an Ojibwa chief of Sault Ste. Marie named Ishquanaby (not translated) and another from Grand Traverse, Eshquagonaby (mistranslated as "the feather of Honor"). There is, however, no evidence that the river or city was named for any individual. Moreover, there is an Escanaba Lake in Vilas County, Wisconsin, but it is probably named for the Michigan city or river.[2]

Goguac Lake and Prairie, on the south edge of the city of Battle Creek, Calhoun County, have a name that has suffered in the hands of the myth weavers. They call it "Indian for pleasant water" and offer Coquaiack and Goguagick as alternative spellings. That interpretation cannot be supported by linguistic evidence from any Michigan tribal language.

It is possible that Goguac is related to the Ojibwa word *goshwakoshka,* "it shakes, it trembles." Even this view is necessarily speculative, since we cannot determine how much the present name is altered from its original form. If our interpretation is valid, it could refer to soft, mucky ground.[3]

The county seat town of Kalkaska and its township are named for their county. In a legislative act of 1843, the county name was spelled Calkaska, but it was changed to Kalkaska in 1871. It has been speculated that the name was manufactured from a combination of Calcraft, for Schoolcraft, and Cass, for the governor. James Calcraft was the great-grandfather of Henry Schoolcraft. It has also been fancied that the name might be derived from an Ojibwa term meaning "burned over." That is the explanation adopted in Kane's *American Counties.*

In Baraga's *Otchipwe Language* the terms that most nearly fit that definition are *kishkakisân (nin),* "I burn through," and *kiskakiswa (nin),* "I burn some object through, in two pieces." The degree of corruption necessary to transform one of these terms into Kalkaska is considerable, but not unprecedented.[4]

Kenockee Township in St. Clair County was organized in 1855. A Kenockee post office also functioned from 1856 to 1878 and 1880 to 1903. Verwyst claimed that its name was from Ojibwa "*ginok,* (he is long-legged), pr. kee-no-kee." That seems unfitting for a place-name. *Gino* does mean long, and *aki* means land; so "long land" is a possible explanation of Kenockee. One could speculate, however, that the name was derived from *kinoje,* "pike, pickerel." Lacking firm evidence, the name should be classed as unsolved.[5]

The chief city of modern Wisconsin, Milwaukee, was once in Michigan Territory, although it was then but a small trading post. Wisconsin Territory was separated from Michigan in 1836. However, the name Milwaukee survives in Michigan, on Milwaukee Lake in Marquette County, Milwaukee Creek, St. Clair County, Milwaukee Junction, a postal station in Detroit and, in corrupted form, Zilwaukee, Saginaw County (see chap. 8). There was once a village of Milwaukie in St. Clair County (1837–1858), named for Milwaukie Creek (presently Milwaukee); it is now Lakeport.

About ten interpretations of this name are in print, although it is not a difficult one to analyze. There can be little doubt that the correct meaning is "good land," from Potawatomi *meno* or *mino,* "good," and *aki,* "land." The *l* sound it now has is the result of misunderstanding by whites.[6]

Nagek Lake in Oceana County appears to have a name shortened from Ojibwa *nagikawad,* "it is nothing."[7]

"Twin Lakes" is the meaning of Nijode Lakes in Mecosta County. The language of this form is Ojibwa.[8]

Quinnesec, a village on the Menominee River in Dickinson County, has its name from Quinnesec Falls in the river, now submerged by a power dam. Writers have interpreted this name as having something to do with smoke or fog, because of the spray from the waterfall. Gagnieur called the name a "contraction of Rakwannessek, a place of smoke or fog," and others have accepted that view, even though there is no *r* sound in the local native languages.

In Ojibwa, *pakwene* or *pashkine* signifies "there is smoke," while *awân* means "it is foggy." There is no approximation of this name in Menominee. No one has explained how the *q* in this name could be derived from the aboriginal *p*. *Q* is not used at all in Baraga, the sound of it being represented by *k*.

A new view is presented here, based on Baraga's *Otchipwe Language: Quinnes* comes from *kiwanis,* "noise," to which is added *ec* for *aki,* "place, ground" or *aka,* "where." "Where it is noisy" would be a suitable translation. Noise is surely as notable an attribute of waterfalls as foggy mist.[9]

The Tacoosh River, Delta County, has a name derived from Ojibwa *takôs,* "short."[10]

The name of Wakeshma Township in Kalamazoo County might be a corruption from Ojibwa *wakeshka,* "it is shining," or a Potawatomi form of the same.[1]

A notable category of Indian place-names refer to the location of a place, although one of Schoolcraft's contributors may have exaggerated slightly when he said "The badge or name of a village is generally taken from the position or place in which it is situated."

Whites, after consulting with the local Ojibwa Indians, named the town and township of Ishpeming, in Marquette County, from the Ojibwa term *ishpiming,* meaning "above, on high, at the top, in the air." The name was also given to Ishpeming Point, the highest place on Isle Royale (el. 1,377 ft.). Ishpeming is only three miles from Negaunee, which has a slightly lower elevation, and so some local wag invented the story that the name Ishpeming meant "Heaven," while Negaunee carried the Ojibwa name for the opposite place. "Ishpeming and Negaunee: Heaven and Hell," the tale went. Indeed Ishpeming is sometimes used as an equivalent for "Heaven," but the Indians knew no hell and had no word for it until Christian influence caused them to represent it with the term *anâmakamig.* Negaunee is from Ojibwa *nigani,* "foremost, ahead, before, in advance, leading." Negani Lake in Iron County, Wisconsin, has its name from the same source.

Negaunee, like Ishpeming, was named by whites. The manner and reason for the choice of names was explained by pioneer Peter White:

> Negaunee was founded in 1857 and Ishpeming in 1858. . . . The pioneer furnace for making pig iron from charcoal with the Jackson ore, was building in 1857 near the Jackson mine. A town was growing up around the mine and furnace, and it was decided to give it an Indian name which should be as nearly as possible a translation of the word "pioneer," inasmuch as the town included the first mine opened and the first furnace built. A council was held with the Ojibwa Indians of the vicinity and the name Negaunee was chosen, which

signifies in Ojibwa, "I take the lead." In the following year the settlers of the growing town about the Cleveland and Lake Superior mines, which are situated upon the dividing ridge between the waters which flow into Lake Superior, and those which flow into Lake Michigan, thought fit to give their town the Indian translation of "on the summit." This proved rather difficult to get into one Ojibwa word. Mr. S. P. Ely was chairman of the committee and came to your narrator for consultation with him as a Chippewa expert. Mr. Ely finally put the decision upon me, and I selected Ishpeming, which is a general term in Ojibwa for any remarkable elevation, and is sometimes applied to Heaven itself.[12]

Otsikita Lake, Lapeer County, has the old Indian name of Lake St. Clair, which lies above Detroit. It has also been spelled Otsiketa and Otsiheta. This does not appear to be an Algonquian name and is probably Wyandot. It could be a corruption of *ochsheetau,* "foot," a name given perhaps because of Lake St. Clair's perceived shape, or because of its location at the foot of the upper lakes system of waterways.[13]

Wauban Beach in Cheboygan County takes its name from *wâban,* the Ojibwa name for "east." It doubles also as the name for the dawn.[14]

Names representing color are reasonably common. Vermilion, treated elsewhere, is an especially abundant name because of the use of vermilion clay for body and facial paint. Its aboriginal name was Oulamon, Onaman, or Osanaman.

Lake Macatawa at Holland in Ottawa County discharges into Lake Michigan between the villages of Macatawa and Ottawa Beach. Whites in early days called it Black Lake. Its name is Ottawa and Potawatomi for "black" and was probably given because of the dark color of the water, caused by tannin from decayed vegetation. The Black River and the Macatawa River, formerly called North Black, flow into it. Blackbird's Ottawa term for "black," inanimate form, is *maw-kaw-te-waw.* Mukutawa Lake and River in the Canadian province of Manitoba have their name from the Cree word for "black."

Another Black River discharges into Lake Michigan at South Haven, and still another joins Lake Superior in Gogebic County. The first was called Rivière Noire, probably in translation from the

Indian name, on Franquelin's map of 1688. The second is shown as "*Muck a tay Sepe, Black R.*" in one of Bela Hubbard's maps of 1840.[15]

The name of Miscauna Creek in Menominee County is probably derived from *miskwa,* "red" (Ojibwa), although the ending, *na,* is not explainable. Baraga's word for "bloody" is *miskwiwan.*[16]

The name of Ossawinamakee Beach, in Schoolcraft County, appears to be from the Ojibwa term signifying "yellow painted land." The location of the name, however, suggests that it was taken from the name of a chief of the Mackinac bands of Chippewa Indians, *O saw waw ne me ke,* who signed a treaty at Detroit on July 31, 1855.[17]

The White River, which empties into Lake Michigan through White Lake in Muskegon County, has a name that is translated from an old Indian name. The river is shown as "*Ouabisipi ou R. Blanche*" (White River) on Franquelin's map of 1688.

This may be the possible origin of the name of the Owasippe Scout Camps, operated by the Chicago Council of the Boy Scouts of America in Muskegon County near Whitehall. They have had their own branch post office, originally called Owasippi, since 1924. The name, which has also been attached to a lake, is attributed to an Ottawa Indian chief of whom many romantic tales are told. About fifty years ago the Scouts even placed a marker on the alleged gravesite of the chief, near the headwaters of Silver Creek, a tributary of the White River.

However, a search of history, treaties, and legends of the Ottawa uncovers no Chief Owasippe. There is a record of an Indian named O-wa-sa-po living in an Ojibwa village at Shiawassee who requested the services of a government blacksmith in 1837. He must have been a very obscure individual, for his name is not found in any treaties. His location is also a long way from the places named Owasippe. We suggest that the native name of the White River, *Waba-sippi* (*Ouabisippi* in French orthography) was turned into Owasippe, and the myth of the chief came next.[18]

## Notes

1. Baraga, *Otchipwe Language* 2:5.
2. Romig, *Michigan Place Names,* 186, 199; Kappler, *Indian Treaties,* 451, 455; Verwyst, "Geographical Names," 391; R. A. Brotherton, "Meaning of Esca-

naba," *Inland Seas* 4 (Fall, 1948): 210–11; Baraga, *Otchipwe Language* 1:104, 214; 2:114; Hubbard, map 4, p. 45, in Peters, *Lake Superior Journal;* Vogel, *Indian Place Names,* 12–13. Prof. Bernard Peters, on the basis of the Kidder manuscript from Chief "Kobawgam," believes that Oshqua na be konk on the Hubbard map signifies "Slippery Rock." Bernard Peters, "The Origin and Meaning of Chippewa Place Names along the Lake Superior Shoreline," *Names* 32 (September, 1984): 239, 242.

3. Romig, *Michigan Place Names,* 227; Albert F. Butler, "Rediscovering Michigan's Prairies," *Michigan History* 32 (March, 1948): 27; Philander Prescott, in Schoolcraft, *Indian Tribes* 2:171.
4. Osborn and Osborn, *Schoolcraft, Longfellow, Hiawatha,* 363, 371; "Reports of Counties," 212; Kane, *American Counties,* 205; Baraga, *Otchipwe Language* 2:191.
5. Romig, *Michigan Place Names,* 301; Verwyst, "Geographical Names," 392; Baraga, *Otchipwe Language* 2:191.
6. Romig, *Michigan Place Names,* 315; Kuhm, "Indian Place-Names," 63–66; Gailland, "English-Potawatomi," 133; Pokagon, *Ogimawkwe Mitigwaki,* 96, 99.
7. Baraga, *Otchipwe Language* 2:266.
8. Ibid., 289.
9. Gagnieur, "Indian Place Names" (1918), 544; see also Kelton, *Indian Names of Places,* 47–48; FWP, *Michigan,* 545; Romig, *Michigan Place Names,* 463; for Menominee, see Hoffman, "Menomini Indians," 295–328; for Ojibwa, Baraga, *Otchipwe Language* 1:106, 154, 181; 2:23, 25, 198.
10. Baraga, *Otchipwe Language* 2:376. Larry Matrious, a Hannahville Potawatomi, believes that the name of Tacoosh River is a perversion of a Potawatomi word meaning "sick." Interview, August 22, 1984.
11. Baraga, *Otchipwe Language* 2:397.
12. Ibid. 1:132; 2:160, 286–87; Verwyst, "Geographical Names," 394; interview with Father John Hascall, pastor at Father Baraga's old parish at L'Anse, Keweenaw Bay Indian Community, August, 1973. Father Hascall is a member of the Bay Mills Chippewa band; Peter White, "The Iron Region of Lake Superior," *MPHSC* 8 (1885): 158–59.
13. Mrs. B. C. Farrand, "Early Days in Desmond and Vicinity," *MPHSC* 13 (1888): 334; Schoolcraft, *Notes on the Iroquois,* 394.
14. Baraga, *Otchipwe Language* 2:390.
15. John T. Blois, *Gazetteer of the State of Michigan* (Detroit: Sydney L. Rood & Co., 1838; reprint, New York: Arno Press, 1975), 255; Romig, *Michigan Place Names,* 343; Blackbird, *History of the Ottawa and Chippewa,* 113; Geographic Board, *Place-Names of Manitoba,* 60.
16. Baraga, *Otchipwe Language* 1:31, 207.
17. Ibid., 2:334; Kappler, *Indian Treaties,* 730.
18. Tucker, *Indian Villages,* pl. 11-B; Romig, *Michigan Place Names,* 424; Emmert, "The Indians of Shiawassee County," 266.

# XVII
# Commemorative Names

Commemorative names recalling historical events or persons were not commonly given by Indians. Names commemorating individuals were virtually unknown before acculturation began. Point Iroquois on Lake Superior (see *Tribal Names*), commemorating a battle between the Iroquois and the Chippewa, is one of the very few commemorative names given by Indians in Michigan. The remaining few names of this class are in English or are Indian words adopted into English, and one of them, Battle Creek, could be a translation from an older Indian name that may in fact survive on one of the tributaries of Battle Creek.

There are at least two stories about the origin of the name of Battle Creek, a tributary of the Kalamazoo River for which the city and township of Battle Creek in Calhoun County are named. The white story is that two Indians and two members of John Mullett's surveying party had a skirmish here in 1824, after which Mullett named the stream Battle Creek.

According to a story allegedly told to an early settler by a Potawatomi chief, Battle Creek received its name in this way:

> Many years ago when there were no she-mo-ko-men [white men] in the country and the red men were plenty and strong, two tribes of his people were at enmity, and many braves, from different tribes, were on the war path. Here, at the junction of these streams [Battle Creek and the Kalamazoo River], and extending up and along the Battle creek and the adjacent country, a mighty battle was fought, and many braves were sent to the happy hunting grounds, for the stream was filled with the dead and its waters were colored with their blood. Hence the name Waupakisco, river of battles or water of blood. . . .

When the city charter was obtained, the name Waupakisco was put to a popular vote and voted down in favor of Battle Creek.

The alleged aboriginal name has nothing to do with blood or battle. It could be from *wabashkiki,* "swamp" (see Wabash). The rest is folk etymology. Wabascan Creek, a tributary of the Kalamazoo River, has a name resembling the supposed early name of Battle Creek, Waupakisco. It does not appear to translate as Battle Creek, however. It does resemble the Ojibwa term *wabishkaan* (*nin*), "I whiten it." The territory was Potawatomi, but the language difference is small.[1]

On a tributary of the Detroit River called Parent's Creek (from a French personal name), on the night of July 31, 1763, a column of 250 English soldiers commanded by Captain Dalzell marched in search of a band of Ojibwa and Ottawa warriors led by Pontiac. They were ambushed near this stream, and when they finally retreated to the fort next morning, they had lost 59 men. Ever after, the stream was called Bloody Run. It still runs through Detroit's Elmwood Cemetery, after which it disappears into the sewer system.[2]

Little Girl's Point, on the Lake Superior shore in Gogebic County, is named from the tradition that an Ojibwa Indian girl drowned at that place.[3] It was shown as *Pt. Petite fille* on Bela Hubbard's map number 16 in 1840.

On an island in the Saginaw River pioneers found a large number of Indian burials. According to Indian tradition, these belonged to Sauk killed in battle by the Ojibwa and their allies in the early eighteenth century. The timing, at least, is dubious, since at that time the Sauk were on the west side of Lake Michigan. Whatever the truth, whites gave the name Skull Island to this place in Bay County.[4]

Numerous commemorative Indian names fit into other name categories, and a number of them are treated under other chapter headings. This includes nearly all of the personal and tribal names, and most places named "Squaw" or "Indian." Squaw Island in Oakland County and Yankee Springs in Barry County are among the notable commemorative names described elsewhere. Virtually all of them were given by whites, but their presence is a measure of Indian influence on language and place-names.

## Notes

1. Romig, *Michigan Place Names,* 46; A. D. P. Van Buren, "The First Settlers in the Township of Battle Creek," *MPHSC* 5 (1882): 292–93; O. Poppleton,

"How Battle Creek Received its Name," *MPHSC* 6 (1883): 248–51; Baraga, *Otchipwe Language* 2:391, 393.

2. Parkman, *Conspiracy of Pontiac* 1:320–29; Ferris E. Lewis, *Michigan Yesterday and Today*, 6th ed. (Hillsdale, Mich.: Hillsdale Educational Publishers, 1967), 124.
3. McKenney, *Sketches*, 223.
4. Albert Miller, "Incidents of Early Saginaw," *MPHSC* 13 (1889): 376–79.

# XVIII

# Names Borrowed from Other States

A large proportion of the aboriginal place-names in Michigan were introduced from the East, mainly from New York. They are thus of dual significance, being a record not only of native influence but also of white settlement patterns. Most of these names are from the New York Iroquois, some are from the Delawares, and a few are from smaller Algonquian tribes of the Northeast. A very few were introduced from the far West and indicate interest in happenings in that region, such as the gold rush.

The purchase of Alaska from Russia in 1867 inspired the adoption of that name for a village in Kent County, December 4, 1868, and for Alaska Bay in Huron County. The name of our fiftieth state is from an Aleut word broadly translated as "the mainland."[1]

Amboy Township in Hillsdale County is named from Perth Amboy, New Jersey. Perth is a Scottish name, while Amboy is a corruption from a Delaware Indian term. According to Heckewelder, it was derived from *emboli,* "round, hollow," while others trace it to *ompoge,* "standing" or "upright."[2]

In the lower Mississippi valley, *bayou* is a generic term for a sluggish stream. It is the French form of the Choctaw *bok,* or *bayuk.* The word is common in eastern Texas, Louisiana, Mississippi, and Arkansas. To a lesser extent it is found in other gulf states and as far north as southern Illinois, and even in Waupaca County, Wisconsin. Suprisingly, there are at least twenty-three instances of it in the Lower Peninsula of Michigan, seventeen of them in Manistee, Ottawa, and Mason counties.

Among examples of *bayou* in Michigan are Indian Pete Bayou and South Bayou on Hamilton Lake, Mason County; Pottawatomie Bayou, a tributary of the Grand River in Ottawa County; and Cisco Bayou on the White River, Oceana County. These bayous are usually stagnant backwaters or bays connected to other bodies of water.[3]

Canandaigua in Lenawee County was named by settlers for a lake and town in Ontario County, New York. The name is from the Seneca language and signifies "chosen town" or "town set off."[4]

Casco is the name of townships in Allegan and St. Clair counties. The second was reportedly named in 1849 for Casco, Maine, by a native of that state, Captain John Clarke. The other was named five years later, but the reason is unknown. The name of Casco Bay, Cumberland County, Maine, is reported to signify "muddy" in the Penobscot language.[5]

Lake Chemung, in Livingston County, is undoubtedly named for Chemung River of New York, or for the county and village bearing the name. It is from an archaic Delaware term meaning "place of the horn" and was given to the river because of the discovery of mastodon tusks along its banks.

A Potawatomi of Huron River named Chamung signed the Treaty of Greenville in 1795. It seems unlikely that the lake was named for him, however, because of the early date of the treaty and because it is in an area where borrowed New York names are numerous.[6]

Chenango Lake and the village named for it in Livingston County draw their name from a town, county, and river in New York. According to Beauchamp, "Che-nan-go is called O-che-nang or *bull thistles* by Morgan and the Onondagas."[7]

The village of Chicora in Allegan County is named for a passenger steamer that sank in Lake Michigan off South Haven, January 21, 1895, with the loss of twenty-six lives. Chicora is a name given in 1521 by Spanish explorers to Indians inhabiting the coast of South Carolina. It is believed to be derived from *Shakori,* the name of a native tribe.[8]

The villages of Cohoctah and Cohoctah Center, as well as Cohoctah Lake and Township, in Livingston County, have names resulting, in all likelihood, from the misspelling of Cohocton, the name of a river in Steuben and Livingston counties, New York, and of a town in Steuben County. Lewis H. Morgan linked it with a Tuscarora word meaning "log in the water."[9]

Coloma, a town and township in Berrien County, have a name inspired by the California gold rush. Coloma, Michigan, was named in 1855 by Gilson Osgood for a village in California where he joined the gold rush in 1849. Coloma, in Eldorado County, California, developed around Sutter's Mill after the discovery of gold during the previous year. It was the name of a village of the Maidu Indian tribe, but its meaning is unknown.[10]

The village of Croton on the Muskegon River in Newaygo County was named in 1850 by the first postmaster, John V. Stearns, for the Croton water works along the Hudson River in Westchester County, New York. It is said that the topography of the Michigan place reminded him of the other. The name has been extended to Croton Dam Pond and to the township. Croton is a name attached to several towns and topographic features above Ossining, New York. Somewhat uncertainly, it has been suggested that it is from Mohican *kenotin,* "the wind."[11]

Cuyahoga Creek and Peak in the Porcupine Mountains of Ontonagon County are named for the Cuyahoga River, which empties into Lake Erie at Cleveland, Ohio. The name is Iroquoian, probably from the Wyandot language, and is said to mean "crooked river," an apt descriptive name for the Ohio stream. However, other explanations are in print.[12]

Genesee County and the village and township of Genesee within it were so named because many of the early settlers came from Genesee County, New York. The New York county is named for its river, which discharges into Lake Ontario at Rochester. The name is from the Seneca, *jonéshi:yoh,* and is generally interpreted as "beautiful valley." The name was also given to a prairie in Kalamazoo County by settlers from New York.[13]

Honeyoey Creek in Oscoda County probably has its name from Honeoye, a town in Ontario County, New York, or Honeoye Lake and River, in that county, or Honeoye Falls, in Monroe County, New York. In Michigan the name is somewhat scrambled but this is a common occurrence. The name is reported to be of Seneca origin (*ha'-ne-a-yah* or *hah'-nyah-yah'*), meaning "finger lying." The story is that an Indian picking strawberries near Honeoye Lake, New York, was bitten by a rattlesnake, so he cut off his finger with his tomahawk and left it lying there.[14]

The village and township of Juniata in Tuscola County are named from a river and county in Pennsylvania. The name is reported to come from a Seneca term, *tyunayate,* "projecting rock," referring to a stone held in reverence by the Indians.[15]

Kearsarge, in Houghton County, was settled in 1867 and later became a copper-mining center, but declined when the mineral was exhausted in 1925. The place was named for the United States navy ship Kearsarge by a former naval officer who became an employee of the Calumet and Hecla Consolidated Copper Company. The Kearsarge played a prominent part in the Civil War. It was the flagship of Admiral David Farragut in the attack on Mobile Bay, August 5, 1864.

The ship was named for one or both of two mountains in the White Mountains of New Hampshire, South Kearsarge in Merrimac County, and North Kearsarge in Carroll County. According to J. H. Trumbull, the name disguises the Pennacook Indian words *koowass-adchu,* "pine mountain." He added, however, that a map of 1784 showed South Kearsarge as "Kyasarga Mountain: by the Indians Cowissawaschook." In this form, Trumbull wrote, it was possible that the name meant "pointed" or "peaked mountain."[16]

Kentucky Lake in Houghton County is named for the state of Kentucky, for which several strange explanations are in print. It does not mean "dark and bloody ground." The best evidence is that it is Wyandot-Iroquoian for "plain" or "meadow land." It is notable that the Mohawk name for LaPrairie, Quebec, is Kentake.[17]

Lehigh Creek in Gogebic County has a name borrowed from Pennsylvania, where it is the name of a river tributary to the Delaware at Easton as well as of a county, state park, university, and several villages and towns. It is a borrowed name in several states. Lehigh is from the Delaware word for "forked stream," *lechauhanne,* given also, according to Heckewelder, to the angle or wedge of land between the confluence of two streams. *Lechau* was also an element in the Delaware name for a forked trail, *lechawekink.* Heckewelder linked the word *lechauhanne* to the junction of the Lehigh River with the Delaware at Easton, while others connected it with the junction of Pahapoco Creek and the Lehigh River near present Lehighton, Pennsylvania. As a generic name, it could be applied to any junction.[18]

The name of Mattawan, a town in Van Buren County, is one variation of an Algonquian term for the junction of two streams. It is found in at least eight states. Baraga called *matawan* a Cree word meaning "it opens (a river), it arrives in a lake." Actually, its application is broader than that. Since Mattawan, Michigan, is not near any stream or lake junction, it is probably a transfer name—perhaps from Matteawan, New York, or Matawan, New Jersey.

Wherever it came from, the name is from an Algonquian language and signifies the same, a meeting of waters. Whites have invented a similar name in Watersmeet, Gogebic County.[19]

Minneapolis Shoal, Delta County, is a borrowed bilingual name, from the city in Minnesota. It is composed of *minne,* the Dakota word for "water," and the Greek *polis,* "city." The city's name was chosen by a local editor in 1852; the *minne* is from Minnehaha Creek, which flows through the city to the Mississippi.

Also borrowed from the Dakota (or Sioux) Indians are the names of Lake Minnewasca in Delta County and Minnewaukon Lake in St. Joseph County. Minnewasca is a spelling variation, or corruption, of a name that appears as Minnewaska on a lake in Pope County, Minnesota, and Minnewasta on a lake in Day County, South Dakota. The last is most accurate and signifies "good water." Minnewaukon is Dakota for "spirit water" and was their name for present Spirit Lake, Iowa. Both names were introduced into Michigan by whites.[20]

The ultimate origin of the name of Norwalk in Manistee County is Norwalk in Fairfield County, Connecticut. It is the modern form of an old Indian (Mohican?) name recorded in colonial times as *Norwaake, Norwauke,* and *Norwaack.* According to J. H. Trumbull, "The fact that the modern spelling of the name was not generally adopted for ten years after the purchase and settlement of the town, is sufficient reason for rejecting the traditional derivation from the day's 'north walk,' to which the bounds of the plantation were extended, from the sea. The name seems to be the equivalent of *Nayaug, Noyack, Nyack,* etc., 'a point of land'."[21]

Onota Township in Alger County near Munising has an obscure name. It was once the name of a village that has all but disappeared. Onota Lake in Berkshire County, Massachusetts, is one possible source of this name. Onoto is an old name for Nanticoke Creek in Broome County, New York, mentioned April 2, 1737, by Conrad Weiser. It has been speculated that the name of the Massachusetts lake might be from a Mohican word for "blue" or "deep." One guess on the New York name is that it is from the Onondaga word *onotes,* "deep."

It is possible that the name in Michigan is a misspelling of Oneota, from the title of Schoolcraft's book, *Oneota, or Characteristics of the Red Race in America* (1845). Therein it is merely one variant spelling of Oneida, the name of one of the six Iroquois tribes, which Schoolcraft said was taken from a sacred stone called *O-ne-a-ta.* Schoolcraft also listed *onontah* as a Wyandot word for hill or mountain.[22]

Oregon Township in Lapeer County has its name from the 1840s when Oregon was much in the public consciousness. The dispute over this name is complex and spirited. It is my view that its ultimate origin is from the Cree *ouragan,* a "dish, plate, or wooden bowl." Along with several other Cree and Ojibwa terms, it probably reached the Northwest via some Indian or French-Canadian employee of the Hudson's Bay Company. The name was applied to the Columbia River as early as 1765, perhaps for the same reason that the Platte River of Nebraska was named by Indians and French for its flat, spread-out character. Other views, some of them adopted in respectable sources, are too involved to recount here, but the curious may examine the works named in the notes.[23]

Otisco Township in Ionia County was named for Otisco in Onondaga County, New York. The first postmaster at the former village of Otisco, Michigan, George E. Dickinson, came from New York. The name is apparently Onondaga for "low water."

Otsego is a name of New York Iroquoian origin that has been given to several places in Michigan. The town of Otsego in Allegan County was named in 1835 by its first postmaster, Horace Comstock, for his home county in New York. The name was extended to the township. Otsego County was also named, in 1843, for the county in New York. In it is Otsego Lake, from which a village, township, and state park have taken the same name. The name Otsego is related to *Otesaga,* given by Lewis H. Morgan as the Mohawk name for Cooperstown, New York. He called its "signification lost," but others translated it as "place of the rock."[24]

Panola Plains in Iron County is a pleasant sounding name, but out of place in Michigan. Panola is found in several places in the South and is the Choctaw word for cotton. Panola is the name of a county in Mississippi and of towns in Alabama, Louisiana, Oklahoma, and Texas. A town called Panola was incorporated in Illinois in 1867, and was apparently named at the suggestion of Civil War veterans. That is the most probable explanation also for its presence in Michigan.[25]

Pewee Lake in Schoolcraft County is named from a colloquial term meaning "very small." The standard works either give no origin for this word, or they disagree with one another. Webster's

*New World Dictionary of the American Language* (1st ed., 1953) called *pee-wee* a Massachuset word meaning "little, small, diminutive, tiny," but gave no origin in the second edition, 1978 printing. There is strong evidence of the correctness of their earlier attribution in Trumbull's *Natick Dictionary*, which is of the Massachuset dialect. It lists "*peawe*, it is small." New England and Virginia Indian languages generated many words in common use, and *peewee* may be one of them. Outside Michigan, we have found three examples of it as a place-name: Pee Wee Hollow in Allegheny County, Virginia; Pewee Run in Rockingham County, Virginia; and Pewee Valley, a town in Oldham County, Kentucky.

A small marble used by children is called a pee-wee, and the name has also been attached to short cowboy boots as well as to a small species of bird. The latter is probably an onomatopoeic name, from the sound of its call. Some have linked this word to penis and to urination, but it is unlikely that anyone would choose a name with such connotations for a place of residence.

*Pewee* is an element of at least one compound place-name, the Pewagen Branch of the Pemmaquon River, Washington County, Maine, which is Malecite for "small portage." The evidence favors Indian origin for peewee.[26]

Podunk Lake, Barry County, has an authentic Indian name of New England origin that continues today as the name of a small river that enters the Connecticut above East Hartford. There it was once the name of a small Indian tribe, probably named from their place of residence, which disappeared from history after King Philip's War in 1676. There is also an unincorporated community called Podunk near Brookfield, Massachusetts, and a Podunk Brook in Washington County, New York. Several other place-names are apparently only spelling variations, including Potunk, in Suffolk County, Long Island, New York; Potick Creek in Greene County, New York; and Potuck Creek in Orange County, New York.

There is some disagreement about the meaning of Podunk. Its origin is apparently in Mohican or related Algonquian languages and dialects of lower New England and New York. Some have associated it with "round," but others see it as a generic term for "a sunken meadow, a wet, miry place, a bog."

Podunk entered the language, according to Allen Walker Read, as a result of an anonymous series of eight letters published in the *Daily National Pilot* of Buffalo, New York, beginning January 5,

1846. Called "Letters from Podunk," they were humorous observations on small town life and were repeated in other papers. Another story says that the name was popularized by vaudeville comedian George M. Cohan, who spent his summers at Podunk, Massachusetts. The Podunk place-names, which spread to Michigan, Wisconsin, Iowa, and Utah, apparently antedate Cohan.[27]

Sandusky in Sanilac County is named for Sandusky, Ohio. Two Wyandot villages, Lower and Upper Sandusky, once existed at the sites of the present cities of Sandusky and Upper Sandusky, Ohio. One of the earliest appearances of the name is Lac Sandouske (now Sandusky Bay) in DeLisle's map of 1718. In 1749 the English established on Sandusky Bay of Lake Erie a trading post called Fort Sandoski, which the Indians burned in 1763. Nearly all interpretations of this name link it with the Wyandot word for water or cool water.[28]

It is said that the town of Saranac in Ionia County was named for the New York resort town in order to attract settlers. In New York, the name is on a lake and town in Franklin County, a town in Clinton County, and a river that joins Lake Champlain at Plattsburgh. There is no certain translation of this name. Beauchamp believed it was "but part of the original name, the terminal of which, *sarrane,* means to ascend." The location of the Saranac names indicates Mohican origin.[29]

Saugatuck, along Lake Michigan in Allegan County, is appropriately named. Its name means "mouth of a river," and it is at the mouth of the Kalamazoo River. However, the name is not from the Potawatomi language, as some have indicated, but from the language of the Mohican of Connecticut or one of the small tribes related to them. This name is obviously a transfer from Saugatuck, Connecticut, which lies at the mouth of a river of the same name. Another Saugatuck River is on Long Island.

The second postmaster at Saugatuck, Michigan (then called Newark) was William G. Butler, from Hartford, Connecticut (1838–1842). According to Romig, it was his successor, Stephen A. Morrison, who was responsible for the adoption of the present name, which was extended to East Saugatuck and to Saugatuck Township.[30]

Sciota Township in Shiawassee County is named from Ohio's Scioto River (so spelled), which runs through Columbus and joins the Ohio River at Portsmouth, Scioto County. The name is from the Wyandot term for "deer."[31]

Shawmut was the native name for the peninsula on which downtown Boston now stands, and the name is still borne by a prominent bank in that city. In Houghton County, Michigan, a copper mine was called Shawmut, and from it a creek was named. Shawmut Hills is a community in Kent County. This name is doubtless the equivalent of Shawomet, a Wampanoag village name in the town of Somerset, Bristol County, Massachusetts. *Shawmut* is said to signify "at the neck" and *shawomet,* "a neck of land." Other views are in print, but this explanation best fits the circumstances of the places to which the name was originally applied.[32]

In 1804 the Russians established a post on Alaska's Baranof Island, naming it New Archangel. It was the capital of Russian America until the American purchase of 1867. Americans gave it the Tlingit name, Sitka, and it remained the territorial capital until 1900. The name is said to signify "by the sea" or "on Shi," the native name for Baranof Island. Sitka is one of several names adopted in the lower states as a result of the Alaska purchase and the gold rush of 1898. In Michigan the name Sitka was given to a lake in Alger County and a locality in Newaygo County.[33] (See also Alaska, Klondike.)

A locale in Lake County is called Skookum. This is a word from the Chinook jargon, a former polyglot trade language of the Northwest coast. It was borrowed from the Chehalis Indian term meaning "demon" or "strong." Rapids in a river were called *skookum chuck,* "strong water." Skookum or Skukum appears in place-names of the Northwest Pacific states, British Columbia, and Alaska, and in a few places outside that region. There is a Skookum Lake in Galbraith Township, Algoma District of Ontario.[34]

Sodus, a village in Berrien County, has its name from New York, where it is on Sodus Bay of Lake Ontario, and three towns in Wayne County: Sodus, Sodus Center, and Sodus Point. Morgan wrote the name *Se-o-dose'* and called it a Seneca word of lost significance. Beauchamp found that this name was written as *Aserotus* in the 1770s but said it "has not been well defined."

Sonoma and Sonoma Lake in Calhoun County were named from a town and county in California north of San Francisco. According to A. L. Kroeber, the name is a Wappo Indian suffix meaning "village of."[35]

The chief religious ceremony of some Indian tribes of the western plains was, and still is for some, the Sun Dance. It is appropriate that a town on the former site of such ceremonies in

Wyoming is called Sun Dance. But there is no local reason for the name of Sun Dance Lake in Gogebic County, Michigan. We have not been able to learn why it was adopted, but the abundance of introduced Indian names in this county suggests that a single individual, perhaps from the United States Forest Service, is responsible.[36]

Tacoma Lake in Montcalm County and Lake Tacoma in Wayne County are clearly named for the city in Washington, which in turn was named for the indigenous name of the mountain now bearing the name of a British naval officer, Mt. Rainier. The meaning of Tacoma is disputed. It is a coast Salish word, which some say comes from *tahoma,* generic for "mountain," while others say it means "nourishing breasts," from the milky appearance of the glacial melt water that streams down from the peak.[37]

Lake Taho in Clare County is an apparent borrowing of the name of Lake Tahoe, situated in the Sierra Nevada on the California-Nevada boundary. It is the generic name for "lake" in the language of the Washo Indians of that region.[38]

Tallahassee Creek in Branch County is named for the capital of Florida. The Florida city has its name from an old Seminole town that formerly stood on the site. It is derived from Creek *talwa,* "town," and *hasi,* "old"—hence, "Old Town."[39]

Texas Creek in Ontonagon County and Texas Township in Kalamazoo County are among the places named during the excitement of the Texas war of independence and the Mexican War ten years later. *Texas* was a term meaning "friends" or "allies" and was used as a greeting by the Hasinai, a Caddoan tribe of the lower Rio Grande region.[40]

Tioga River in Baraga County and Tioga Lake in Livingston County have the same name as a river of New York and Pennsylvania and of counties and villages in those states. The name was originally given to an Indian village at the junction of the Chemung and Susquehanna rivers, near present Athens, Pennsylvania. According to Heckewelder, the name meant "a gate, a place of entrance" (from *tiaóga*) because a trail leading from the Delaware to the Iroquois country began at that point. Lewis H. Morgan linked the name with the Cayuga term *ta-ya-o-ga,* meaning "at the forks," referring to the convergence of several trails at that place. Actually, however, the name was also applied to the junction of streams, and there is such a junction at each of the places in New York and Pennsylvania where the name occurs.[41]

Tonawanda Lake in Grand Traverse County has a Seneca Indian name from western New York. There it is attached to several places in Erie, Niagara, and Genesee counties, as well as to an Indian reservation. It is translated as "swift water."[42]

The village and township of Unadilla in Livingston County are named for a town in Otsego County, New York. *Unadilla* is said to be an Oneida word meaning "place of meeting." It has also been borrowed as a name for places in Georgia and Nebraska.[43]

Wabash Creek is a tributary of the Two Hearted River in Luce County. It is named for the Wabash River of Indiana and is the only example of the occurrence of this name outside Indiana and Illinois. Jacob P. Dunn, a scholar of the Miami Indian language, relying on information supplied by Gabriel Godfroy, a Miami speaker, established that this name was abbreviated from *Wahbahshik-ki*, meaning "white stone river."[44]

*Wahoo* is a word with several uses. It can be a meaningless yell. It is the common name of a shrubby western tree, also called arrowwood or burning bush, of which there are two species, *Euonymus atropurpureus* and *Euonymus americanus*. From that source, it is believed, comes the name of Wahoo, in Saunders County, Nebraska. W. R. Gerard held that this word was from Dakota *wahoo*, which, with the first vowel nasalized, meant "arrowwood." J. P. Williamson's *English-Dakota Dictionary* lists *wanhi* (nasalized) for "arrowhead." The name Wahoo has also been borne by a southern tree of the elm family (*Ulmus alata*), and is drawn from the Muskhogean word *uhahwu*, of unknown meaning. Finally, *wahoo* is the name of a species of salt water fish, *Acanthocybium solanderi*.

All the above is an introduction to a name in Michigan, Wahoo Prairie Drain, in Monroe and Lenawee County. The best guess is that it is named for the shrub otherwise called burning bush, which is not native to Michigan but is widely planted as an ornamental.[45]

Winnipeg Lake in Calhoun County is named for the very large lake of the same name in Manitoba, Canada. The name is from Ojibwa *winipeg* or Cree *we'*nipak (or *winnipek*), meaning "salt water" or "unclean water." It is their name for the sea and was transferred to large bodies of fresh water even if, as in the case of Lake Winnipeg, the water was neither unclean nor salty.[46]

In the early eighteenth century, along the Susquehanna River at the present site of Wilkes-Barre, Pennsylvania, stood a village of Delaware Indians. They called the settlement and the valley about it M'*chewomink,* "upon the great plains." Whites wrote it as *Chiwaumuc, Waiomink, Wajomik, Wiawamic,* and finally, *Wyoming.* So was born a famous name, which was later to be given to a great western state and to nineteen other places, including the city of Wyoming in Kent County, Michigan.

The name was not famous until after a battle and a poem made it so. The battle took place during the Revolutionary War, July 3, 1778, when white settlers who had crowded out the Delaware inhabitants were attacked by a mixed force of eleven hundred men, consisting of two hundred British soldiers, two hundred Tory supporters, and seven hundred Iroquois Indians, led by Major John Butler.

Enough settlers were killed by the attackers to earn for this event the name Wyoming massacre. Thirty-one years later, in 1809, a Scottish poet, Thomas Campbell, published at London his *Gertrude of Wyoming, a Pennsylvania Tale, and Other Poems.* The heroine was fictional, but incidents of the tragedy were not. Campbell's book became popular, and soon the name Wyoming was being adopted for towns and townships, counties, and topographic features from Rhode Island to the far West, where, in 1869, it was given to a state. It was even adopted as a town name in Ontario.

Wyoming Township, Michigan, for which the city in Kent County, a suburb of Grand Rapids, was named, was organized in 1848, and clearly had no connection with the state in the West but rather was commemorative of the now nearly forgotten valley in Pennsylvania.[47]

Yuba Creek in Grand Traverse County is one of several names adopted in Michigan as a result of the California gold rush. It was near a Maidu Indian village called Yuba that gold was first discovered in California in 1848. From it was named the river at the mouth of which it was located, and in 1850 Yuba County was created. The old Indian village site is today occupied by Yuba City, seat of Sutter County, on the west side of the Yuba River. The meaning of Yuba is undetermined.[48] (See also Coloma, Sonoma.)

"The main source for American place names," wrote one names scholar, "appears to be place names already on the map elsewhere." Perhaps this phenomenon has some value as an indicator of the places from which settlers came, or what events elsewhere were influencing their thoughts, but another writer has concluded that "imported names lose their interest etymologically as soon as they are traced to another locality."[49]

## Notes

1. Romig, *Michigan Place Names,* 14; J. Ellis Ransom, "Derivation of the Word Alaska," *American Anthropologist,* n.s., 42 (1940): 550–51.
2. John G. Heckewelder, *Narrative of the Mission of the United Brethren among the Delaware and Mohegan Indians* (Cleveland: Burrows Brothers, 1907), 560; Ruttenber, *Footprints,* 102.
3. Byington, *Choctaw Language,* 94; Georg Friederici, *Amerikanistisches Wörterbuch* (Hamburg: Cram, De Gruyter & Co., 1960), 84–85; Read, *Louisiana Place Names,* xii; Romig, *Michigan Place Names,* 523–24; correspondence, Richard Kasperson, Northbrook, Illinois, April 19, 1983.
4. Romig, *Michigan Place Names,* 96; Beauchamp, *Aboriginal Place Names,* 155–56.
5. Romig, *Michigan Place Names,* 101; Fannie H. Eckstorm, *Indian Place-Names of the Penobscot Valley and Maine Coast* (Orono, Maine: University Press, 1941), 169.
6. Beauchamp, *Aboriginal Place Names,* 41–42; Kappler, *Indian Treaties,* 44.
7. Beauchamp, *Aboriginal Place Names,* 44.
8. Hodge, *Handbook of American Indians* 1:263; John R. Swanton, *Indian Tribes of North America,* Bureau of American Ethnology Bulletin no. 145 (Washington, D.C.: U.S. Government Printing Office, 1952), 83–84; FWP, *Michigan,* 128; Romig, *Michigan Place Names,* 114.
9. Beauchamp, *Aboriginal Place Names,* 104.
10. Fox, "Place Names of Berrien County," 21; Powers, *Tribes of California,* 315.
11. Romig, *Michigan Place Names,* 141; Beauchamp, *Aboriginal Place Names,* 52, 176.
12. Martin, "Ohio Place Names," 277; August C. Mahr, "Indian River and Place Names in Ohio," *Ohio Historical Quarterly* 66 (1957): 154.
13. Romig, *Michigan Place Names,* 220; Chafe, *Seneca Morphology,* 68, no. 1118.
14. Beauchamp, *Aboriginal Place Names,* 157.
15. Donehoo, *Indian Villages,* 75.
16. Romig, *Michigan Place Names,* 298–99; Trumbull, "Indian Geographical Names," 20.
17. John P. Harrington, *Our State Names* (Washington, D.C.: Smithsonian Institution Press, 1955), 380; Abbe J. W. Forbes in *Kateri* no. 122 (Winter, 1979): 27–28.
18. Heckewelder, *Mission,* appendix, pp. 542–43; Daniel G. Brinton and Albert S. Anthony, *A Lenape-English Dictionary* (Philadelphia: Historical Society of Pennsylvania, 1888), 62; Hodge, *Handbook of American Indians* 1:763; Ruttenber, *Footprints,* 167.
19. Ruttenber, *Footprints,* 37.
20. Upham, *Minnesota Geographic Names,* 233, 432; Virginia Driving Hawk Sneve, *South Dakota Geographic Names* (Sioux Falls: Brevet Press, 1973), 348; Riggs, *Dakota-English,* 317; Vogel, *Iowa Place Names,* 44.
21. James H. Trumbull, *Indian Names in Connecticut* (reprint, Hamden, Conn.: Archon Press, 1974), 40.
22. John C. Huden, *Indian Place Names of New England* (New York: Museum of

the American Indian, 1962), 156; Beauchamp, *Aboriginal Place Names,* 29; Schoolcraft, *Notes on the Iroquois,* 441, 396.

23. Kellogg, *Early Narratives,* 91; Vogel, *Indian Place Names,* 97–98, 176; Virgil Vogel, "Oregon, a Rejoinder," *Names* 16 (June, 1968): 136–40; George R. Stewart, "Ouarican Revisited," *Names* 15 (September, 1967): 166–68; Vernon F. Snow, "From Ouragon to Oregon," *Oregon Historical Quarterly* 60 (December, 1959): 439–47.
24. Romig, *Michigan Place Names,* 303, 422; Beauchamp, *Aboriginal Place Names,* 148–49, 174; Morgan, *League of the Iroquois* 2:138.
25. William A. Read, *Indian Place-Names in Alabama* (Baton Rouge: Louisiana State University Press, 1937), 51; Vogel, *Indian Place Names,* 104; Byington, *Choctaw Language,* 319.
26. Trumbull, *Natick Dictionary,* 323; Partridge, *A Dictionary of Slang,* 479, 491; Huden, *Indian Place Names,* 185.
27. Romig, *Michigan Place Names,* 449; Trumbull, *Indian Names,* 52; Hodge, *Handbook of American Indians* 2:271; Beauchamp, *Aboriginal Place Names,* 85, 164, 220; William W. Tooker, *The Indian Place Names on Long Island* (reprint, Port Washington, N.Y.: Ira J. Friedman, 1962), 197; Huden, *Indian Place Names,* 189; Allen W. Read, "The Rationale of Podunk," *American Speech* 14 (April, 1939): 99–108; *Chicago Daily News,* Beeline, January 23, 1978.
28. Vogel, *Indian Place Names,* 122–23; Hodge, *Handbook of American Indians* 2:431.
29. Romig, *Michigan Place Names,* 500; Beauchamp, *Aboriginal Place Names,* 45.
30. Trumbull, *Indian Names,* 64; Beauchamp, *Aboriginal Place Names,* 223; Romig, *Michigan Place Names,* 172, 501.
31. Martin, "Ohio Place Names," 276; Mahr, "Indian River and Place Names," 140; Reuben G. Thwaites, ed., "The French Regime in Wisconsin," *Collections State Historical Society of Wisconsin* 18 (1908): 20n.27.
32. Clarence Barnhart, ed., *New Century Cyclopedia of Names* (New York: Appleton Century Crofts, 1954), 3:3578; R. A. Douglas-Lithgow, *A Dictionary of American-Indian Place and Proper Names in New England* (Salem, Mass.: Salem Press, 1909), 160; Huden, *Indian Place Names,* 229.
33. Donald J. Orth, *Dictionary of Alaska Place Names,* U.S. Geological Survey Professional Paper no. 567 (Washington, D.C.: U.S. Government Printing Office, 1967), 880.
34. Gibbs, *Chinook Jargon,* 23.
35. Sodus: Morgan, *League of the Iroquois* 2:132; Beauchamp, *Aboriginal Place Names,* 36, 241. Sonoma: A. L. Kroeber, *Handbook of the Indians of California* (reprint, New York: Dover Publications, 1976), 897.
36. Ruth M. Underhill, *Red Man's Religion* (Chicago: University of Chicago Press, 1965), chap. 14.
37. Edmond S. Meany, *Origin of Washington Geographic Names* (Seattle: University of Washington Press, 1923; reprint, Detroit: Gale Research Co., 1968), 299–300; Myron Eells, "Aboriginal Geographic Names in the State of Washington," *American Anthropologist* 5 (January, 1892): 31–33.
38. Kroeber, *Indians of California,* 897; Helen S. Carlson, *Nevada Place Names* (Reno: University of Nevada Press, 1974), 228–29.

39. William A. Read, *Florida Place Names of Indian Origin* (Baton Rouge: Louisiana State University Press, 1934), 33.
40. H. E. Bolton in Hodge, *Handbook of American Indians* 2:738–41.
41. Heckewelder, *Mission,* 555–56; Morgan, *League of the Iroquois* 2:102, 133.
42. Beauchamp, *Aboriginal Place Names,* 83.
43. Ibid., 44.
44. Jacob P. Dunn, "Indiana Geographical Nomenclature," *Indiana Quarterly Magazine of History* 8 (September, 1912): 113–14.
45. *Webster's New World Dictionary,* 1596; Lilian L. Fitzpatrick, *Nebraska Place Names* (Lincoln: University of Nebraska Press, 1960), 128; Gerard, "Plant Names," 303; John P. Williamson, *An English-Dakota Dictionary* (Minneapolis: Ross & Haines, 1970), 8.
46. Baraga, *Otchipwe Language* 1:301; Watkins, *Dictionary of the Cree,* 163.
47. Swanton, *Indian Tribes,* 54; Heckewelder, *Mission,* 449; Hodge, *Handbook of American Indians* 2:978–79; Donehoo, *Indian Villages,* 259–63.
48. Erwin Gudde, *1000 California Place Names* (Berkeley: University of California Press, 1959), 96.
49. John Rydjord, *Kansas Place-Names* (Norman: University of Oklahoma Press, 1972), 249; William F. Ganong, "A Monograph on the Place-nomenclature of the Province of New Brunswick," *Royal Society of Canada, Proceedings and Transactions* (1896), 183.

# XIX

# Indian Names from Outside the United States

Michigan Indian names from other countries of the Americas—Canada, the Caribbean, and Latin-American nations—were all introduced by whites.

At various times, at least six settlements in Michigan have borne the name Canada or Canadian. At least two of them were named by or for settlers from Canada. A rural post office named Canada Creek operated in Montmorency County from 1939 to 1941. The United States Geological Survey still lists as populated places Canada Corners, Muskegon County, and Canada Shores, Branch County. Vanished names of nineteenth-century settlements are Canada, St. Joseph County, and two places called Canada Settlement, in Eaton and Ionia counties. Several water features are named for Canada: Canada Creek in Ontonagon County, Canada Creek, rising in Montmorency County and joining the Black River on the line of Presque Isle and Cheboygan counties, and Canada Lake, in Clare County.

The name Canada first appears in the accounts of the expeditions of Jacques Cartier along the St. Lawrence in 1534–36. It was the name for a "kingdom" about the site of the present city of Quebec, then called Stadacona, reaching from the island of Orleans below the city to a point about ten miles upstream. "The Lord of Canada," Cartier wrote, was Donnaconna, and "the Canadians with eight or nine villages alongst the river be subjects unto them." In his glossary, Canada was defined as "a Towne."

Father Baraga defined Canada as an Iroquois word for "a village of tents or huts." Heckewelder was probably correct in guessing that

> it is highly probable that the Frenchman who first asked the Indians in Canada the name of their country, pointing to the

> spot and to the objects surrounding him, received for answer *Kanada,* (town or village), and . . . believed it to be the name of the whole region, and reported it so to his countrymen, and consequently gave to their newly acquired dominion the name of Canada.[1]

The story of the origin of Canada is one of many instances in which the name of a local area has been extended to a country or even a continent.

*Canoe,* the name given by whites to the dugouts or bark- or skin-covered boats of the Indians, originated in the language of the West Indian Arawaks. Columbus borrowed the word from the Haitians, and so it was adopted into Spanish, French, and English.

Canoe Rocks off Isle Royale in Lake Superior are named for their appearance. Canoe Lake, in Schoolcraft County, likewise appears to be named for its shape.[2] Other canoe names are Canoe Bay, Lake, and Point in Chippewa County and Canoe Highway, a channel in St. Clair County.

Several names from Latin America that were adopted in Michigan were probably inspired by literature. The town of Capac in St. Clair County, according to the Federal Writers Program, "was named, no one knows why, for Manco Capac, traditional founder of the Inca dynasty." Schoolcraft, commenting on his legendary story of Iosco in *Algic Researches,* related it to Capac: "The story itself, so far as respects the object, is calculated to remind the reader of South American History, of the alleged descent of Manco Capac and the children of the Sun."

The dates of the rule of Manco Capac are unknown. Two later rulers bore the name—Manïta Capac in the thirteenth century and Huaina Capac, 1493–1527. *Capac* is said to mean "chief" and is derived from the Aymara word *kapac,* "falcon."[3]

The Chocolay River, which gives its name to a township in Marquette County, joins Lake Superior just east of the city of Marquette. Early accounts called it the Chocolate River. The present spelling is apparently an attempt to convey French pronunciation of the name. The stream was probably named for the color of its water, created by tannin from decaying vegetation. *Chocolate* is a word first adopted into Spanish from the Aztec *chocolotl.* A chocolate beverage flavored with vanilla was a favorite drink of the Aztec ruler Moctezuma.[4]

Cougar Lake in Iron County is named for the American panther, also called puma. *Cougar* is a word from the Tupi-Guarani language of Brazil, adopted into English from the Portuguese.[5]

Henry R. Schoolcraft, coasting the south shore of Lake Superior during the summer of 1820 in the company of an expedition led by Governor Lewis Cass, wrote, "We encamped on a beach of sand, near the entrance of a small creek, which, from a violent storm that raged during the night, we called Hurricane Creek." The stream, in eastern Alger County, is now called the Hurricane River. The name *hurricane,* borrowed from that of the whirling tropical storm, is not properly applied to inland storms of North America, but nevertheless it has been given to many topographic features of the United States, owing to storms that occurred in those places.

The word *hurricane* is from the West Indian word *huracan,* early adopted into Spanish but not listed in any English dictionary before 1720. According to Daniel Brinton, the Quiche Mayas of Guatemala adopted the name, as *Hurakan,* for "the mysterious creative power." He and others called it a borrowed word of Haitian origin, and if that is so, it is from the Arawak language, although Bartlett called it a Carib word. The Caribs were a people with a distinct language, but when their name is used in a geographic sense for the Caribbean area, confusion results, since a people of different speech, the Arawaks, occupied the larger West Indian islands.[6]

Kamloops Island, Keweenaw County, has the name of an Indian tribe and city of British Columbia. The name has also been given to a species of salmon. The city began as a Hudson's Bay Company post in 1812. The native Indians of the Shuswap-Salish stock are the source of the name, which has been interpreted as "confluence," for the meeting of the North and South Thompson rivers at that place. Others have explained it as "point between the rivers," while some Indians reportedly said it meant simply "meeting place."[7]

Klondike Creek in Alger County and Klondike Lake in Schoolcraft County bear one of the several names from the Alaska-Yukon region that were adopted elsewhere as a result of the gold rush that peaked in 1898. Klondike is the name of a small tributary of the Yukon River in Yukon Territory, Canada, along which an important gold strike was made, causing a rush from the port of Skagway into the interior. The only agreement about this name is that it is corrupted from a local Athabascan name, written in many ways, that has been interpreted as "deer," "deer river," and "hammer water"—from a native custom of pounding sticks into shallow streams to support salmon nets. It may have something to do with fish, from its similarity to the names of nearby Kloo Lake (Fish Lake) and Kluane Lake (*Kloo-ah-nee*), "Large Fish Lake."[8]

The Peruvian Incas had a god named Rimac, whose temple "was resorted to by countless numbers from all parts of the realm." The name was from the Quechua tongue and signifies "he who speaks." The name Rimac is preserved today on a Peruvian river, but the Spanish corrupted it to Lima when they named their colonial capital in 1535. Lima Township in Washtenaw County is named for the Peruvian capital.[9]

Oronoko Township in Berrien County was named in 1837 by Governor Stevens T. Mason, according to one writer, "in honor of Oronoko, the Indian chief." But there was no such chief. Another writer speculates that the name is of Chippewa origin, but that is also erroneous. The name is obviously only a variant spelling of Orinoco, the name of the South American river that discharges into the Caribbean through Venezuela. The name has been adopted in three other states besides Michigan, and all spell it differently: Oronoco, Minnesota, Oronogo, Missouri, and Ornonoque, Kansas.

It is possible that the name in Michigan originated in Thomas Southern's popular play, *Orinoko* (1696), which was followed by various imitations and adaptations during the eighteenth century. The name was borrowed from a novel, *Oroonoko; or, the Royal Slave,* by Aphra Behn, published in London in 1688. In both the novel and the play, the hero is a Negro of Surinam. There can be little doubt that Mrs. Behn borrowed the name from that of the river.

It is also possible that the name in Michigan came from a variety of strong tobacco mentioned in early accounts as *oronoco, orinoko,* and *oronooka. Orinoco* is reported to be a Carib word meaning "clay river."[10]

The Potato River in Ontonagon County and Potato Creek in Saginaw County are probably not named for the cultivated potato but for edible wild tubers sometimes called Indian potatoes. The arrowhead (*Sagittaria latifolia*) and Jerusalem artichoke (*Helianthus tuberosus*) are among several plants so described by early writers.[11]

According to Dustin, Potato Creek in Saginaw County was called *Wau-po Se-be* by the Ojibwa Indians. This may be corrupted from *Wabisipin Sibi,* "swan's potato river." The arrowhead plant was called "swan's potato" by many Indians, and Wapsipinicon River in Iowa was named for it.

Standing alone, the Ojibwa word for "potato" is *opin.* Potato is a word that has evolved from *batata,* applied by the Taino (Arawakan) Indians of Haiti to the sweet potato, a member of the morning glory family. The Spanish extended it to the Andean tuber of the nightshade family that is now called potato in several European languages. True potatoes were cultivated in Peru in pre-Columbian times and were destined to become a major part of the world's food supply.[12]

*Tobacco* is a word so thoroughly naturalized into English that its native origin is forgotten. Alexander von Humboldt reported (1804), "The word tobacco (tabacco) . . . belongs to the ancient language of Hayti, or St. Domingo. It did not properly denote the herb, but the tube through which the smoke was inhaled." The Tupi-Guarani (Brazilian) name for this plant was *petun,* from which came our word *petunia.* Several varieties of wild tobacco grew in North America, and some were cultivated by Indians. A mixture of smoking herbs and bark used by the Ojibwa and other Indians was called *kinnikinnick* (mixture), which is used as a place-name in some states. Tobacco as a place-name came indirectly from the Caribbean by way of Spanish and English. Michigan has two Tobacco rivers. One, a tributary of the Tittibawassee River in Gladwin County, has given its name to a township; the other joins Lake Superior in Keweenaw County. It is reported that the latter stream was named for the brown color of its water.[13]

North Americans are the most assiduous name borrowers on earth. On the other hand, no North American names are found south of the Rio Grande.

## Notes

1. Romig, *Michigan Place Names,* 96; Henry S. Burrage, ed., *Early English and French Voyages 1534–1608* (New York: Charles Scribner's, 1906; reprint,

New York: Dover Publications, n.d.), 44–47, 60, 88; Baraga, *Otchipwe Language* 1:298; Heckewelder, letter to DuPonceau, September 5, 1816, in *History of the Indian Nations,* 422.

2. Wilfred Funk, *Word Origins and their Romantic Stories* (New York: Wilfred Funk, 1950), 346; Dustin, "Isle Royale Place Names," 696.
3. FWP, *Michigan,* 439; Williams, *Schoolcraft's Indian Legends,* 146; Alain Gheerbrant, ed., *The Incas, Royal Commentaries of the Inca Garcilaso de la Vega* (New York: Orion Press, 1961), 418, 420.
4. Chocolate River, so named in Article I, Chippewa Treaty of October 4, 1842, in Kappler, *Indian Treaties,* 542; Eric Partridge, *Origins: A Short Etymological Dictionary of Modern English* (London: Routledge & Kegan Paul, 1958), 96; William H. Prescott, *History of the Conquest of Mexico and History of the Conquest of Peru* (New York: Modern Library, n.d.), 79, 323.
5. Walter W. Skeat, *Etymological Dictionary of the English Language* (Oxford: Clarendon Press, 1911; 4th ed., London: Oxford University Press, 1961), 138.
6. Henry R. Schoolcraft, *Narrative Journal . . . in the Year 1820* (Albany: E. & F. Hosford, 1821), 149; cited by Bernard Peters (1981); Bartlett, *Dictionary of Americanisms,* 209; Daniel G. Brinton, *The Myths of the New World,* 2d ed. (New York: Henry Holt, 1876), 52; Walter W. Skeat, *The Language of Mexico, and Words of West Indian Origin* (London: Philological Society, 1890), 145.
7. Hamilton, *Canadian Place Names,* 52; Hodge, *Handbook of American Indians* 1:649; G. P. V. Akrigg and Helen Akrigg, *1001 British Columbia Place Names* (Vancouver: Discovery Press, 1973), 93.
8. Hodge, *Handbook of American Indians* 1:714–15; Mathews, *Dictionary of Americanisms,* 938; James W. Phillips, *Alaska-Yukon Place Names* (Seattle: University of Washington Press, 1973), 75–76; Hamilton, *Canadian Place Names,* 322.
9. Brinton, *Myths of the New World,* 32; Prescott, *History of the Conquest,* 1006–7.
10. Romig, *Michigan Place Names,* 419–20; Fox, "Place Names of Berrien County," 27–28; Bissell, *American Indian in English Literature,* 78–84, 140n., 201; Mathews, *Dictionary of Americanisms,* 1069; Nils M. Holmer, "Indian Place Names in South America and the Antilles, Part I," *Names* 8 (September, 1960): 144.
11. Virgil J. Vogel, "American Indian Foods Used as Medicine," in *American Folk Medicine,* ed. Wayland C. Hand (Berkeley: University of California Press, 1976), 133–35.
12. Dustin, "Some Indian Place-names," 733; Baraga, *Otchipwe Language* 2:333, 366, 393; Gerard, "Plant Names," 292.
13. Alexander von Humboldt, *Personal Narrative of Travels to the Equinoctial Regions of America* (London: George Bell & Sons, 1877), 2:566n.; Friederici, *Amerikanistisches Wörterbuch,* 494; Skeat, *Etymological Dictionary,* 776; Chamberlain, "Algonkian Words," 246; Bernard C. Peters, "The Origin of Some Stream Names along Michigan's Lake Superior Shoreline," *Inland Seas* 37 (Spring, 1981): 8, citing Douglass Houghton.

# XX
# French-Indian Names

The French were the first white men to explore Michigan, and they held sway there from the early seventeenth century until they were ousted by the English in the French and Indian War, which ended in 1763. The French continued to be the numerically dominant European nationality in the territory of the future state for more than fifty years thereafter.

They left their mark on the state in the form of numerous place-names, such as Detroit, River Rouge, and Sault Ste. Marie. There are uncounted numbers of such names on obscure villages, creeks, and topographic features. What is not generally known is that many of the French names are translations from Indian names. How many of them are taken from the native toponymy is unknown, for no one has undertaken the large task of tracing all of them. The few that are known and described here are, although not in the native language, a measure of aboriginal influence and deserve to be included in this survey. Some of them, in fact, such as Baie de Wasai, are Franco-Indian combinations.

French-Indian mixed bloods, *métis,* are responsible for many of the names dealt with here. "Geographic science," wrote Louise Houghton, "owes something to the French mixed-bloods, aside from the services which they have rendered to exploration. They have also borne some part in map-making." She pointed out that the French geographer J. N. Nicollet was accompanied by a French-Ojibwa guide in his explorations of the Great Lakes and upper Mississippi in the 1830s, some of which carried him to Michigan's Upper Peninsula. It was due to his guide, Houghton wrote, that Nicollet was able "with such wonderful accuracy" to set down "in his chart many lakes, rivers, creeks and islands which he did not see."[1]

Nestled on the edge of a fragment of the Bay Mills Indian Reservation on Sugar Island, the largest American island along the St. Marys River, in Chippewa County, is the village of Baie de

Wasai. It takes its name from an inlet of Lake Nicolet, which is part of the river. *Wasai* is from the Ojibwa word *awâssi,* translated by Baraga as "Burbot, a freshwater fish." Gagnieur translated the name as "Turbot Bay," and described "turbot" as a bullhead. The *burbot* or *turbot* is a freshwater codfish that has been likened to a cross between an eel and a bullhead. By some it is called "lawyer." The fish is said to abound in the vicinity of the village of Baie de Wasai.[2]

Bois Blanc is the name of a large island in Lake Huron, three miles southeast of Mackinac Island. The island is also a township. Bois Blanc is French for "white wood." The Ojibwa name was *Wikabiminiss* or *Wigobiminiss,* "Basswood island" (*wigobi,* "basswood," plus *miniss,* "island"). The wood of this tree (*Tilia americana*) is white, and the bark of it was used for cordage and wigwam construction. The Algonquian name of this tree, often written as *wicopy,* literally signifies "tying bark," and, Gerard wrote, "The tying-bark par excellence of the aborigines was the bast of the Linden [basswood]. . . ."[3]

Calumet Lake and the town and township of Calumet in Houghton County were named in 1866 for the Calumet and Hecla Mining Company. Calumet is a name found also in Illinois, Indiana, Minnesota, and Canada. It was also adopted for a brand of baking powder. The face of an Indian on the baking powder cans created the impression that this is an Indian name. Some Indians even made the cans into rattles used in dances.

Calumet is actually derived from an old Norman-French word for a shepherd's pipe, but in America it was applied by the French to the Indian peace pipe. In August, 1763, Alexander Henry mentioned reaching, on the Ottawa River of Canada, "the portage du Grand Calumet . . . which name is derived from the pierre à Calumet, or pipestone." Father Marquette, on his first meeting with a band of Illinois Indians along the Mississippi in 1673, was offered a pipe to smoke "as a token of peace." He added that "These pipes for smoking tobacco are called in this country *Calumets.*" There was, he remarked,

> nothing more mysterious or respected among them. Less honor is paid to the crowns and scepters of Kings than the savages bestow upon this. It seems to be the god of peace and of war, the arbiter of life and of death. It has but to be carried upon one's person, and displayed, to enable one to walk safely

> through the midst of enemies, who, in the hottest of the fight, lay down their arms when it is shown. For that reason, the Ilinois [*sic*] gave me one, to serve as a safeguard among all the nations through whom I had to pass during my voyages.[4]

(See also Pipestone, Berrien County.)

Ships going to or from Sault Ste. Marie via the St. Marys River enter or leave Lake Huron via De Tour Passage, between the mainland in Chippewa County and Drummond Island. The town on the point on the west side of the passage is named De Tour Village. Formerly it was De Tour, but the word *village* was added to avoid confusing motorists. De Tour is reportedly a translation of the Ojibwa name *Giwideonaning*, "point where we go around in a canoe."[5]

The Ecorse River, a tributary of the Detroit River, gives its name to the city of Ecorse in Wayne County. The French called the stream *Rivière aux Ecorses*, "bark river," for the bark obtained there for canoes. The French are said to have given the name in translation from the Indian name. The Ojibwa equivalent of this name would be *wigwass* (bark) plus *sibi* (river).[6]

On July 15, 1826, Thomas McKenney, commissioner of Indian affairs, coasting the south shore of Lake Superior with a party led by territorial governor Lewis Cass and Indian agent Henry Schoolcraft, reported:

> We found ourselves near Point au Sablé, which is the commencement of a most extraordinary mountain of sand, called *Grand Sablé*, which varies in height from one hundred to three hundred feet, and stretches along the shore of the lake, at the base of which is a beach of not more than ten feet wide, for *nine miles*.

The Ojibwa name of this place was *Negawadjing*, "At the sand mountain." It is the *Nagow Wudjoo* (sand dunes) of Longfellow's *Hiawatha* poem, and it is here that Pau-puk-keewis danced his beggar's dance at Hiawatha's wedding feast:

> Then along the sandy margin
> Of the lake, the Big-Sea-Water
> On he sped with frenzied gestures

Stamped upon the sand, and tossed it
Wildly in the air around him;
Till the wind became a whirlwind,
Till the sand was blown and sifted
Like great snowdrifts o'er the landscape,
Heaping all the shores with Sand Dunes,
Sand Hills of the Nagow Wudjoo!

The French adopted and translated the Ojibwa name of the sand dunes, which are now within the Pictured Rocks National Lakeshore, just west of Grand Marais, calling the place Grand Sable, from which came the names of Grand Sable Lake and Au Sable Point. Nearby Grand Marais is also a corruption from the French translation of the Ojibwa name *kitchi-bitobig,* "great pond." Mapmakers, according to Professor Bernard Peters, erred in changing the French *mare* (pond) to *marais* (marsh).[7]

When the French voyageurs crossed the mouth of a bay from one headland to another, the short cut was called a *traverse.* From it came several place-names on the American map. The best known of this genre in Michigan are Grand Traverse and Little Traverse bays on Lake Michigan, Grand Traverse County, and Traverse City. Grand Traverse is a modification of the original French appellation, La Grand Traverse, while Little Traverse is an Anglicization of La Petite Traverse. The Ojibwa name for a *traverse* was *niminâgan,* but their name for the bay was Kitchi Wekwetong, while the site of Traverse City was Wequetong, "head of the bay." Their name for Little Traverse Bay was Wequetonsing, "at the head of the small bay." It is preserved in the name of a small village in Emmet County. There are also several Traverse names along the Keweenaw Peninsula: Grand Traverse and Little Traverse bays, Traverse Point, Traverse Island, Traverse River, and Traverse Lake. However, there is no evidence at hand that their names are of native origin.[8]

The town, township, and Indian reservation called L'Anse ("the bay or cove") are named for their location at the foot of Keweenaw Bay. The Ojibwa name for this place was *Wequadong, Wikwedong,* or *Wikwetong,* "end of the bay."

As we have noted earlier in our discussion of material culture names, the name of the Flint River is a translation from the Ojibwa name. The French called it La Pierre, which was corrupted into Lapeer as a county name.[9]

One of the most celebrated French names of Indian origin in Michigan was *L'Arbre Croche* (the crooked tree). No longer on the map as such, it is commemorated today by an unsatisfactory substitute, Cross Village in Emmet County. The present name recalls a cross erected on the site by Jesuit missionaries. According to Andrew J. Blackbird, an Ottawa who spent his life in this region, L'Arbre Croche was an Ottawa village that stood on the site of the present village of Good Hart, but others say that the village of L'Arbre Croche stretched along the Lake Michigan shore for several miles. It was named, Blackbird wrote, "From a crooked pine tree at Middle Village [Good Hart] whose top was crooked, almost hooklike."[10]

Alexander Henry, a captive of the Ottawa, gave a description of this village in 1763:

> At the entrance of Lake Michigan, and at about twenty miles to the west of Fort Michilimackinac, is the village of L'Arbre Croche, inhabited by a band of Otawas [*sic*], boasting of two hundred and fifty fighting men. L'Arbre Croche is the seat of the Jesuit mission of Saint Ignace de Michilimackinac, and the people are partly baptized, and partly not. The missionary resides on a farm, attached to the mission, and situated between the village and the fort, both of which are under his care. The Otawas of L'Arbre Croche, who, when compared with the Chipeways, appear to be much advanced in civilization, grow maize for the market of Michilimackinac, where this community is depended upon, for provisioning the canoes.[11]

The Ottawa equivalent of the name L'Arbre Croche was given by Blackbird as *Waw-gaw-naw-ke-zee,* by John Tanner as *War-gun-uk-ke-zee,* and by Henry Schoolcraft as *Wakanukkizzie.*[12]

Les Cheneaux Islands ("the channels") are in Lake Huron, just offshore in eastern Mackinac County. One author says the name is a translation from the "Indian" name *Shebawononing,* meaning "channels." The Ojibwa name for a channel is *inâonan,* a corruption of which, *awanon,* can be seen in the above name. *Sheb* is harder to explain, but may be from *sibi,* "river." The terminal *ing* is locative. Dwight Kelton called Les Cheneaux a translation from the Ojibwa name but gave its original form as *Anaminang,* "in the bowels," from *anamina,* "underneath," or "in the body." He held that the name referred "to the intricate tortuosity of the channels."[13]

In Mackinac County are Millecoquin Lake, the village of Millecoquins, the Millecoquin River, and Millecoquins Point. Derived from this name by corruption is the name of Milakokia Lake. The meaning in French is "a thousand thieves." However, Father Gagnieur claimed that these names are distortions of the Ojibwa *mananakoking* or *minaking,* signifying "place where the hardwood, especially the iron-wood, is plentiful."[14]

Point aux Chenes, "Oak Point," in Mackinac County, is a name reportedly translated by the French from the Ojibwa *nemitigomishking.*

Seul Choix Point ("only choice") in Schoolcraft County is a name supposed to originate in a story that sailors caught in a storm took shelter there because it was their "only choice." Father Gagnieur, however, maintained that those who lived there called the place *Shishewah,* and he traced this to the Ojibwa word *shashoweg,* said to refer to the "straight line" of the coast. Presumably, this was corrupted into Seul Choix.[15]

"The renown of the French," wrote Bacqueville de La Potherie in the eighteenth century, "was made known in the most remote countries." The proof of it can be seen in the names they left.[16]

## Notes

1. Louise S. Houghton, *Our Debt to the Red Men: The French-Indians in the Development of the United States* (Boston: Stratford Co., 1918), 128–29.
2. Baraga, *Otchipwe Language* 1:37; Gagnieur, "Indian Place Names" (1918), 535; Romig, *Michigan Place Names,* 39.
3. Gagnieur, "Indian Place Names" (1918), 554; Baraga, *Otchipwe Language* 1:23, 28; 2:242, 414; Gerard, "Plant Names," 303.
4. Lahontan, *New Voyages* 1:402; Armour, *Attack at Michilimackinac,* 12; Kellogg, *Early Narratives,* 239, 244–45; Gagnieur, "Indian Place Names" (1918), 540–42; see also Blackbird, *History of the Ottawa and Chippewa,* 95.
5. Gagnieur, "Indian Place Names" (1918), 533.
6. FWP, *Michigan,* 469; Baraga, *Otchipwe Language* 1:23, 214.
7. McKenney, *Sketches,* 183–84; FWP, *Michigan,* 561; Gagnieur, "Indian Place Names" (1918), 539; Bernard C. Peters, "The Origin and Meaning of Place Names along Pictured Rocks National Lakeshore," *Michigan Academician* 14 (Summer, 1981): 41–43.

8. "Appointment of Augustin Hamelin Jr. as head chief by Ottawas," *MPHSC* 2 (1887): 621–22; Baraga, *Otchipwe Language* 1:24, 268; Elvin L. Sprague and Mrs. George N. Smith, *Sprague's History of Grand Traverse and Leelanau Counties* (n.p.: B. F. Bowen, 1903), 219; Romig, *Michigan Place Names,* 590; Verwyst, "Geographical Names," 398.
9. Warren, *Ojibway Nation,* 86; Gagnieur, "Indian Place Names" (1918), 540; Jenks, "County Names" (1926), 651.
10. Blackbird, *History of the Ottawa and Chippewa,* 10, 37; Romig, *Michigan Place Names,* 141.
11. Henry, *Travels and Adventures,* 47.
12. Blackbird, *History of the Ottawa and Chippewa,* 10; Tanner, *Narrative of Captivity,* 19; Schoolcraft, *Indian Tribes* 2:31.
13. Romig, *Michigan Place Names,* 325; Kelton, *Indian Names of Places,* 33.
14. Gagnieur, "Indian Place Names" (1918), 552.
15. Ibid., 552–53.
16. Blair, *Indian Tribes* 1:348.

# XXI
# Potpourri

There are names on the map that defy explanation. Some of these have an aboriginal ring, but on closer scrutiny they turn out to be acronyms (such as Wahjamega) or corruptions of words from French or other European languages. Others are from native terms that have become too mangled to be analyzed. It is possible that further research could explain some of them. What follows is informed speculation.

Aginaw Lake, Shiawassee County, appears to have the name of Saginaw minus the *s*.

Gokee Creek in Cheboygan County could have its name from Ojibwa *gogi*, "dive"—i.e., "diving creek."[1]

Kinwamakwad Lake in Gogebic County might have its name corrupted from *ginwakwad*, Ojibwa for "wood," in the inanimate form.[2]

The name Meauwataka on a village and lake in Wexford County has been interpreted to mean "halfway." However, Baraga's word for halfway, *abitawikana*, has little resemblance to this. The name cannot now be explained.[3]

Misteguay Creek is a tributary of the Bad River in Saginaw County. On one old map it was reportedly labeled *Michtegavock*. It has been suggested that the name "is a corruption of 'Me-zhe-say,' wild turkey, together with some qualifying word." The Ojibwa word for turkey as given by Baraga is *misisse*. One could guess that the prefix *mis*, or *mich*, signifies "great," while *teguay* is a corruption of *tigweia*, an alternate term for "river." However, this stream is not a great river, and *tigweia* is normally used in connection with a verb. The meaning of this name is presently unsolved.[4]

Big and Little Muscamoot bays and Muscamoot Ridge (island) in Lake St. Clair, St. Clair County, could have received their name by corruption from *Muckamoot*, the name of a Potawatomi chief who, with his followers, was arrested and held at Owosso in 1840, to be transported to a western reservation.[5]

A long-shot guess on the origin of the name of Neahtawanta Point, Grand Traverse County, is that it is from Ojibwa *nawaiiwan,* "in the middle, center, between."[6]

Nunica, a village in Ottawa County, has a name said by Romig to be "derived from the Indian menonica, meaning clay earth, from which they made pottery." However, the Ojibwa word for "clay" is *wabigan,* and nothing like *menonica* appears in Ojibwa language sources examined. There is little chance that the name in this location comes from another Indian language.[7]

The name Ocqueoc in Presque Isle County is shrouded in mystery. It was given first to a river and then transferred to a lake, a waterfall, and a tiny village. The two explanations in print, "crooked water" and "sacred water," do not have the slightest foundation in Ojibwa vocabularies, although this was their territory. We could guess that the name is a corruption from *Ocanauk,* an Ojibwa signer of the Treaty of Saginaw, September 24, 1819. This could be the same individual whose name is written *Oc-que-wi-Sank* in a memorial to Congress, October 3, 1832, asking that terms of the Treaty of 1819 be honored. In Baraga's *Otchipwe Language* the terms most nearly approximating this name are *okwik,* "cross stick in a snowshoe," and *okwanim,* "beaver dam." Possibly the river was not named for an individual, but for some feature.

On January 18, 1984, Virginia R. Schaedig of Ocqueoc provided the following information from a history of Ocqueoc that was written by her daughter, Cindy S. Schaedig, while a senior in college:

> Though all but one family of Indians had left Ocqueoc by the time the settlers moved in, they were the ones who gave Ocqueoc its name. George Cannon did the first survey of Ocqueoc in 1855–56. In his field notes he recorded the name as Wauk-Wa-Auk. This was most likely his attempt to phonetically spell what he heard the Indians call this area. Fred Denny Larke, the editor of the Presque Isle Advance, recorded the original name of Ocqueoc [as] Waw-Waugh-Waugh-Que-Oc. He put this in the paper in 1878 but did not cite a source for his information. "Crooked River" or "Crooked Waters" is the meaning given to Wauk-Wa-Auk [citing Romig, *Michigan Place Names*]. This name is probably of Huron Indian origin. The third name for Ocqueoc, We-

Qwe-Og, refers to Hammond Bay rather than the river itself and is of Chippewa Indian origin.

Mrs. Schaedig adds, "The source of the last name was an old Chippewa Indian still living in the area." *Waga* or *wagi* is a root meaning "crooked," and sometimes appears as *wauka,* as in Waukazoo, discussed elsewhere. One of the names given by Cindy Schaedig, Wauk-Wa-Auk, could conceivably signify "crooked tree," although corrupted. The name is clearly not from the Huron language. In its present form, it has no meaning.[8]

Omena Bay and Point and the village of Omena in Leelanau County have an obscure name. In 1852 the Reverend Peter Dougherty established a mission to the Ottawa there that continued for nearly twenty years. Dougherty was the first postmaster at this place, and local lore claims that his response to any statement made by an Indian was "Omenah?" (Is it so). The name has also been erroneously interpreted as "a point beyond." However, Father Verwyst offered a more likely explanation. He called the name a corruption of *o minan* (he gives to him), which was pronounced *o-meenan.* In Baraga's *Otchipwe Language* is the phrase *ni nmina,* "I give him."[9] The name Omena appears again on a lake in St. Joseph County.

A good guess on the origin of the name of Onega Lake in Schoolcraft County is that it comes from *Oneaga,* the Cayuga name for the Niagara River, as given by Morgan.[10] He gave its supposed meaning as "at the neck." Schoolcraft could have been the name giver in this case, as several Iroquoian names in Michigan originated with him.

The Quanicassee River, a tributary of Saginaw Bay, and the village at its mouth, in Tuscola County, have an obscure name. Although sources define it as "lone tree," that view is incompatible with vocabularies of Ojibwa and Ottawa. There is a chance that this name is a corruption from the name of Aughquanahquosa (Stump-tail Bear), an Ottawa who signed a treaty at Greenville, Ohio, July 22, 1814, or O-quai-naasa, an Ottawa (possibly the same person), who signed a treaty at Maumee Rapids, Ohio, on August 30, 1831.[11]

Wanadoga Creek, a tributary of Battle Creek in Calhoun County, has a name of unknown origin. Its ending, *doga,* appears to be Iroquoian. Conceivably it is a corruption of Onondaga, perhaps introduced by immigrants from New York. Onondaga in Ingham County was named in that way.

Wau Me Ga Lake in Oakland County has a name similar to that of Wamegan, an Indian who was killed by a bear near Saginaw. However, there is nothing to tie the lake name to him.[12]

Wico, a settlement in Gogebic County, could have its name from *wigob,* often written *wicopy,* the Ojibwa name for the basswood tree, which was much used for cordage and in dwelling construction.[13] (See Bois Blanc in the discussion of French-Indian names.)

The name of Winnewanna Lake in Washtenaw County could be corrupted from *minnewawa,* a word used in Longfellow's *Hiawatha* to represent "a pleasant sound, as of wind in the trees." It is the name of a waterfall in Ontonagon County. (See also Minnewanna Lake, Lapeer County.)

Winnepesaug Lake in Oceana County may have its name shortened from that of Winnepesaukee Lake in New Hampshire. The name is spelled in many ways and has been attributed to the Pennacook, but interpretations vary.[14]

The following names are aboriginal in appearance, but there is not enough information about them to warrant a guess as to their origin and meaning: Amaung Lake in Newaygo County; Ankedosh Creek in Chippewa County; Chippeny Creek in Delta County; Lake Co Be Ac in Roscommon County (possibly an acronym); Copneconic Lake in Genesee County; Kewagawan Lake in Kent County; Meximinee Falls in Gogebic County; Musketeep Lake in Oceana County; Ponshewaing, a settlement in Emmet County; Sawkaw Lake in Newaygo County; Shinahguag Lake in Genesee County; Lake Skegemog in Kalkaska County; Squaconning Creek in Bay County; Webinquaw Lake in Oceana County; and Lake Winnogene in Manistee County.

This may be the appropriate place to quote a verse by Eva March Tappan:

> We drove the Indians out of the land,
> But a dire revenge these redmen planned,
> For they fastened a name on every nook,
> And every boy with a spelling book
> Will have to toil till his hair turns gray
> Before he can spell them the proper way.[15]

To which we might add, scholars must scramble their brains to figure out their origin and meaning.

## Notes

1. Baraga, *Otchipwe Language* 2:141.
2. Ibid., 162.
3. Romig, *Michigan Place Names,* 360; Baraga, *Otchipwe Language* 1:129.
4. Dustin, "Some Indian Place-names," 733; Baraga, *Otchipwe Language* 1:270; 2:387.
5. Niles, "Old Times in Clinton County," 625; the *White Pigeon Republican* of August 28, 1839, printed a summary of a speech made by "Muckmota" in council at "Notawassippi" on August 20, 1839, in which the chief strongly opposed removal of Indians to the West; reprinted in *MPHSC* 10 (1887): 171–72.
6. Baraga, *Otchipwe Language* 2:279.
7. Romig, *Michigan Place Names,* 408; Baraga, *Otchipwe Language* 1:50.
8. Romig, *Michigan Place Names,* 412; FWP, *Michigan,* 483; Kappler, *Indian Treaties,* 187; Carter, *Michigan, 1829–37,* 529; Baraga, *Otchipwe Language* 2:322; Mrs. Schaedig's material is slightly edited.
9. Virgil J. Vogel, "The Missionary as Acculturation Agent: Peter Dougherty and the Indians of Grand Traverse," *Michigan History* 51 (Fall, 1967): 198–201; Romig, *Michigan Place Names,* 414; Wilson, "George N. Smith," 206; Verwyst, "Geographical Names," 395; Baraga, *Otchipwe Language* 1:115.
10. Morgan, *League of the Iroquois* 2:chart following p. 60.
11. FWP, *Michigan,* 434; Romig, *Michigan Place Names,* 462; Kappler, *Indian Treaties,* 106, 339.
12. Williams, "Personal Reminiscences," 240.
13. Baraga, *Otchipwe Language* 2:414; Chamberlain, "Algonkian Words," 266.
14. Huden, *Indian Place Names,* 7, 289.
15. Tappan, "On the Cape," stanza 1, quoted in *Bartlett's Quotations.*

# XXII
# Michigan Indians Today

The number of American Indians in Michigan is not easy to determine. Most observers believe that they are undercounted by the Bureau of the Census. The 1980 census reported that Michigan had 40,038 Indian residents, making it eighth in rank among the states for Indian population. Indian residents are reported in every county in the state, ranging from 1 person in Keweenaw County to 6,667 in Wayne County, which includes Detroit. The latter figure is questioned by the Michigan Commission on Indian Affairs, which believes there may be as many as 18,000 Indians in Wayne County. Many Indians living in larger cities are from tribes not native to Michigan.

The great majority of Michigan's Indians live off the reservations. Detroit has the largest number of these, while substantial numbers are reported also in Grand Rapids, Lansing, Saginaw, Flint, and Sault Ste. Marie. The Three Fires, consisting of the Ojibwa (Chippewa), Ottawa, and Potawatomi tribes constitute all of the Michigan Indian population still living on reservations or distinct Indian communities.

The largest Indian reservation in Michigan is the Keweenaw Bay or L'Anse Reservation of Ojibwa Indians in Baraga County. It was established by the Treaty of LaPointe in 1854 and consists of about 14,000 acres. The resident population of about five hundred lives principally in two villages, Assinins and Zeba, on opposite sides of Keweenaw Bay. A detached community officially belonging to the Keweenaw Reservation is Indian Village on Lac Vieux Desert in Gogebic County. The ancient name of this place, as earlier indicated, was Katikitegon, "garden place." Here a population of less than a hundred people lives on 80 acres of reserved land.

Near Ontonagon in the county of the same name is the Ontonagon Reservation, which was created in 1854 and now consists of 724 acres. It is deserted, but the 1980 census reported sixty-two Indians in Ontonagon County.

The Bay Mills reservation on Waiska Bay, Chippewa County, consists of 2,139 acres and has a population of three hundred. A detached dependent portion of this reservation, consisting of 600 acres, is at Baie de Wasai on Sugar Island, just east of Sault Ste. Marie. The reservation is named for sawmills once located on Waiska Bay. The Indians of Bay Mills are descendants of those called *Sauvages Saulteurs* (Falls Indians) by the French because they lived along the rapids of St. Marys River at the outlet of Lake Superior. The aboriginal name of this place was *Pawating*, "at the rapids." The abundant supply of fish found there supported a large Indian population. Today it is estimated that a thousand Ojibwa still live in the city of Sault Ste. Marie, where they own 80 acres of land purchased by themselves and held in trust by the Department of the Interior.

At Brimley, Michigan, the Bay Mills Indians operate the Bay Mills School of Business, which is open to all Michigan Indians.

It is estimated that there are fifteen hundred Chippewa and Ottawa Indians without trust land living primarily in Delta, Schoolcraft, Mackinac, and Charlevoix counties.

The only significant Ojibwa reservation in the Lower Peninsula is the Isabella Reservation in Isabella County. It takes its name from the county in which it is situated, which is named for Queen Isabella, the benefactor of Columbus. The reservation, in Chippewa Township, three miles east of Mount Pleasant, was established August 2, 1855, by treaty with the Saginaw, Black River, and Swan Creek bands of Ojibwa Indians. Today it consists of 1,786 acres with a population of about two hundred. Here the tribe operates a small wood products factory. A detached community of about twenty Indians belonging to the Isabella Reservation is at Saganing on Lake Huron, in Arenac County.

On Beaver and Hog islands in Lake Michigan, Charlevoix County, are 1,749 acres of allotted trust lands held by Chippewa and Ottawa Indians. The government recognizes the Grand Traverse band of Chippewa and Ottawa Indians at Peshawbestown, Leelanau County (mailing address: Suttons Bay).

The Ottawa Indians, as a tribe, no longer possess a contiguous reservation in Michigan, but do own scattered tracts. Many Ottawa were removed to Kansas and later Oklahoma in the last century, while others fled to Canada, where their descendants live on Manitoulin Island. However, a relatively large number of Ottawa remain in their old homeland about Grand Traverse and Little Traverse bays. Some are organized into the Northern Michigan Ottawa Association, with headquarters at Petoskey. This organization owns about 4,500 acres of land scattered over a wide area, prosecutes claims before the Indian Claims Commission, and holds annual powwows.

Peshawbestown, on the west arm of Grand Traverse Bay in Leelanau County, is the headquarters of the Grand Traverse Band of Ottawa and Chippewa Indians, with a membership of about a thousand scattered over five counties. Peshawbestown is exclusively an Indian community, with a population in 1984 of about two hundred persons living on individually owned tracts of land. Other Ottawa live in Traverse City, Petoskey, Harbor Springs, Good Hart, Cross Village, and elsewhere.

The Grand River Band of Ottawa was organized in 1948 to prosecute land claims against the United States. Its members are mainly in Grand Rapids and Grand Haven.

The number of Ottawa in Michigan is estimated at more than two thousand.

The Potawatomi of Michigan are found in both Upper and Lower peninsulas. When the Chicago Treaty of 1833 surrendered all Potawatomi lands in Illinois and southern Wisconsin and provided for the removal of the Indians to western Iowa, some fled into northern Wisconsin, where their descendants live on trust lands near Crandon, in Forest County, which were granted them by federal legislation in 1915. Some of these Indians moved farther east into Menominee County, Michigan, where in 1883 the Reverend Peter Marksman, a Methodist missionary, provided land for them and loaned them money to establish themselves. In 1913 Congress provided for the purchase of nearly 3,400 acres of land for them, which was distributed in individual allotments. This tract is called Hannahville Reservation, named for Hannah Marksman, the wife of the Indians' benefactor. Its population in 1969 was estimated at two hundred.

In the southwest part of the Lower Peninsula of Michigan are three Potawatomi communities that contain descendants of two historic bands. The first of these is called the Pokagon Band, named for their early chief, Leopold Pokagon (ca. 1775–1841), and his son, Simon Pokagon (ca. 1830–99). As indicated earlier, the name Pokagon survives on a village in Cass County.

After most of the Potawatomi were removed from Indiana and Michigan to Kansas between 1837 and 1840, a few followers of Pokagon were permitted to remain in Berrien County. Their descendants eventually settled in two other locations, using money from the sale of their old lands to buy new tracts. One group settled at Rush Lake, northwest of Hartford in present Van Buren County, while the others remained near Long Lake in the vicinity of Dowagiac, Cass County. Some live also near Cassopolis. The number of Pokagon Band Potawatomi in Berrien, Cass, and Van Buren counties was estimated at 637 in 1972.

The other group of Potawatomi having descendants in Michigan is called Nottawasippi (variously spelled), from their residence on the river of that name in Calhoun and St. Joseph counties. They were removed to Kansas by military force in 1840, but a number of families trudged back to Michigan, where sympathetic whites helped them buy 120 acres of land near Athens in Calhoun County. This acreage is now held in trust for the Indians by the state of Michigan and is called a state reservation. Their community is locally known as Indian Town. A few other members of this group live at Pine Creek.

Southern Michigan Potawatomi are represented by three organizations: the Potawatomi Indian Nation, Inc., of Hartford, the Potawatomi Indians of Indiana and Michigan, at Dowagiac, and the Huron Potawatomi, Inc., whose chairman lives at Fulton, Kalamazoo County.

As previously indicated, most Michigan Indians are urban dwellers. They are not eligible for federal health care and other benefits enjoyed by those living on federal reservations. Many urban Indians maintain ties with their home reservations and communities and frequently return there to visit.

Urban Indians are well organized, but usually along pan-Indian rather than tribal lines. Their organizations, together with nonreservation Indian groups in rural areas and small towns, have been said to number at least thirty-eight. Among them are the North American Indian Association, operators of the Detroit Indian Center; the Southeastern Michigan Indians, of Centerline, a Detroit suburb; the Grand Valley American Indian Lodge and the Grand Rapids Inter-Tribal Council, Grand Rapids; American Indians Unlimited, Ann Arbor; the Lansing North American Indian Center; and the Intertribal Association, of Saginaw. At Northern

Michigan University in Marquette is the Organization of North American Indian Students, which formerly published a notable monthly paper, the *Nishnawbe News.*

The urban Indian organizations strive to provide a center for social and educational activities, social services, Indian cultural events, and in some instances, political action.[1]

## Notes

1. This information is based on the following sources: Michigan Commission on Indian Affairs, *Michigan Indian Directory* (Lansing: 1981); idem, *Annual Report for 1981* (Lansing: 1982); National Native American Cooperative, *Native American Directory* (San Carlos, Ariz.: 1982); Charles F. Cleland, *A Brief History of Michigan's Indians* (Lansing: Michigan History Division, Michigan Department of State, 1975), 25, 30–35; James Clifton, *The Pokagons, 1683–1983* (Lanham, Md.: University Press of America, 1984); Edmund J. Danziger, *The Chippewas of Lake Superior* (Norman: University of Oklahoma Press, 1978), passim; FWP, *Michigan,* 37, 542, 603; Theodore W. Taylor, *The States and Their Indian Citizens* (Washington, D.C.: U.S. Department of Interior, 1972), 229, 239; United States House of Representatives, *Report with Respect to the House Resolution Authorizing . . . an Investigation of the Bureau of Indian Affairs* (Washington, D.C.: U.S. Government Printing Office, 1953), 759, 761, 1166–67, 1171, 1176, 1178; United States Department of Commerce, *Federal and State Indian Reservations* (Washington, D.C.: U.S. Government Printing Office, 1971), 156–62; Winger, *Potawatomi Indians,* 86–87, 149, 157; *Chicago Daily News,* February 7, 1978; *New York Times,* October 18, 1970, April 3, 1978; personal information from Emelia Schaub of Leland, Michigan, and the following Indian informants: Father John Hascall of Keweenaw Bay, Michigan; Ruth Bussey, Catherine Baldwin, and Ardith Harris of Peshawbestown, Michigan; Arnold Sowmick and Eli Thomas of Isabella Reservation, Mt. Pleasant, Michigan; Sally Eichhorn and Irene Mishigaud of the Hannahville Potawatomi Community.

# Bibliography

## Unpublished Sources


Augustine, Robertson. "Indians, Sawmills, and Danes." Greenville, Mich.: Flat River Historical Society, 1971. Typescript.

Chouteau, August. "Notes on the Indians of North America." MS from *Ancient and Miscellaneous Surveys* 4, St. Louis, February 2, 1816. National Archives.

Elmer, Zindel. "Amerindian Loan Words in American English." Master's thesis, Columbia University, 1955.

Evans, Martha E. "City of White Cloud." June, 1979. Mimeo.

Gailland, Rev. Maurice. "English-Potawatomi Dictionary." Bureau of American Ethnology catalog no. 1761. Ca. 1870. MS, microfilm.

Gogebic County Clerk. "List of Waterfalls in Gogebic County." n.d.

Hartley, Alan H. "Preliminary Observations on Ojibwa Place Names." Duluth, Minnesota, April, 1981. Photocopy of typescript.

Jipson, Norton W. "The Story of the Winnebagoes," and "Winnebago Vocabulary." Chicago Historical Society (1924?).

Kent County Parks and Forests Department. Historical information, n.d. Mimeo.

Le Boullenger, Joseph Ignatius. "French-Illinois Dictionary." 1718. MS. In John Carter Brown Library; photocopy in Newberry Library.

Lemmer, Victor F. "History of the County of Gogebic." 1965. Mimeo.

Marquette County Clerk. "Death Certificate of Charles Kawbawgam," No. 147-564, recorded January 4, 1903. Photocopy.

Marquette County Historical Society. "Ghosts of Iron." List of ghost towns, n.d. Mimeo.

Michigan Historical Commission. "Native Americans of the Historic Period." Typed biographical information, from which 27 profiles were photocopied.

National Archives. "Letters Received by the Office of Indian Affairs, 1824–81. Chicago Agency, 1824-1834." Microcopy M-234, Roll No. 132.

Phelps, J. W. "Diary Kept while at Mackinac, Chicago, and Western Posts, 1840–41." New York Public Library item no. 825, photocopy at Regenstein Library, University of Chicago.

United States Geological Survey, Geographic Names Information System, National Gazetteer of the United States. "Michigan Place Names," 1983. Computer printout.


## Correspondents


Awdey, Mary K. Kent County Park Commission, January 24, 1984.

Bernhardt, Harold. Iron County Historical and Museum Society, January 20, 1984.

Bloomquist, Beatrice. Iron Mountain, Michigan, January 25, 1984.
Carlson, Bruce. City Clerk, Bessemer, Michigan, January 30, 1984.
Durand, Paul. Prior Lake, Minnesota, February 6, 1984.
Heuvelhorst, Patricia. Little Traverse Regional Historical Society, Petoskey, Michigan, August 11, 1982.
Kasperson, Richard. Northbrook, Illinois, April 19, 1983.
Larson, Catherine A. Kalamazoo Public Library, Kalamazoo, Michigan, March 6, 1984.
Le Blanc, William A. Director Michigan Commission on Indian Affairs, May 18, 1983.
Lockhart, Andrew. President Ontonagon County Historical Society, Ontonagon, Michigan, January 30, 1984.
Paull, Frank O., Jr. Director Marquette County Historical Society, Marquette, Michigan, January 13, 1983.
Payne, Roger L. Manager Geographic Names Information System, U.S. Geological Survey, Reston, Virginia, December 7, 1982.
Peters, Bernard. Geography Department, Northern Michigan University, Marquette, Michigan, December 30, 1982.
Peterson, Melvin R. Clerk of Gogebic County, Bessemer, Michigan, January 23, 1984.
Schaedig, Virginia R. Ocqueoc, Michigan, January 18, 1984.
Schaub, Emelia. Director Leelanau Historical Museum, Leland, Michigan, May 1, 11, 1984.
Scott, Margene. Deputy Clerk, Speaker Township, Brown City, Michigan, January 20, 1984.
Seeke, Joyce. Deputy Clerk, Gogebic County, Bessemer, Michigan, January 9, 1984.
Sullivan, Isabella. President Alger County Historical Society, Munising, Michigan, January 20, 1983.
Swanberg, Faye. Munising, Michigan, January 9, 1983, January 25, 1984.
Tuffts, Edith M. President Kalkaska County Historical Society, Kalkaska, Michigan, January 20, 1984.
Voyta, Francis J. Ottawa National Forest, Ironwood, Michigan, February 17, 1984.
Williams, H. O. Museum Director, Flat River Historical Society, Greenville, Michigan, February 28, 1985.
Willman, Charles. Secretary-Treasurer, Ontonagan County Historical Society, Ontonagan, Michigan, October 17, 1984.

## Oral Information

Baldwin, Catherine C. Grand Traverse Band of Ottawa and Chippewa Indians, Peshawbestown, Michigan, June 18, 1984.
Bussey, Ruth. Grand Traverse Band of Ottawa and Chippewa Indians, Peshawbestown, Michigan, June 18, 1984.
Clark, Sandra. Editor, *Michigan History*, August 2, 1982.

Eichhorn, Sally. Hannahville Potawatomi Community, Wilson, Michigan, August 22, 1984.
Halsey, John R. Archaeologist, Michigan History Division, Department of State, Lansing, Michigan, August 2, 1982.
Harris, Ardith (Dodie). Grand Traverse Band of Ottawa and Chippewa Indians, Peshawbestown, Michigan, June 18, 1984.
Hascall, Reverend John. Catholic pastor of Keweenaw Bay Chippewa Community, Assinins, Michigan. Member, Bay Mills Chippewa Band, August 20, 1973.
Matrious, Larry. Hannahville Potawatomi Community, Wilson, Michigan, August 22, 1984.
Mishigaud, Irene. Hannahville Potawatomi Community, Wilson, Michigan, August 22, 1984.
Sowmick, Arnold (Brown Beaver). Chairman, Isabella Band of Chippewas, Mount Pleasant, Michigan, August 3, 1982.
Thomas, Eli (Little Elk). Member, Midewiwin and Isabella Chippewa Band, Mount Pleasant, Michigan, August 3, 1982.

## Dictionaries and Vocabularies

Adams, Ramon. *Western Words: A Dictionary of the American West.* Norman: University of Oklahoma Press, 1968.
Assiniwi, Bernard. *Lexique des Noms Indiens en Amérique.* Ottawa: Editions Lemeac, 1973.
Baraga, Frederic. *A Dictionary of the Otchipwe Language, Explained in English.* 2 vols. in one. Reprint. Minneapolis: Ross & Haines, 1966.
Bartlett, John R. *Dictionary of Americanisms: A Glossary of Words and Phrases . . . Peculiar to the United States.* 3d ed. Boston: Little, Brown, 1860.
Brinton, Daniel G., and Albert S. Anthony. *A Lenape-English Dictionary.* Philadelphia: Historical Society of Pennsylvania, 1888.
Byington, Cyrus. *A Dictionary of the Choctaw Language.* Bureau of American Ethnology Bulletin no. 46. Washington, D.C.: U.S. Government Printing Office, 1915.
Chafe, Wallace. *Seneca Morphology and Dictionary.* Washington, D.C.: Smithsonian Institution Press, 1967.
Cuoq, J. A. *Lexique de la Langue Algonquine.* Montreal: J. Chapleau & Fils, 1886.
Friederici, Georg. *Amerikanistisches Wörterbuch.* Hamburg: Cram, De Gruyter & Co., 1960.
Gibbs, George. *A Dictionary of the Chinook Jargon or Trade Language of Oregon.* New York: Cramoisy Press, 1863.
International Colportage Mission. *A . . . Dictionary of the Ojibwa and English Languages.* Toronto: 1912.
Lahontan, Louis Armand, Baron de. "A Short Dictionary of the Most Universal Language of the Savages." In *New Voyages to North-America.* Edited by R. G. Thwaites, vol. 2:733–48. Chicago: A. C. McClurg, 1905.

Lemoine, George. *Dictionnaire Francaise-Montagnais.* Boston: W. B. Cabot & P. Cabot, 1901.
Mathews, Mitford W. *Dictionary of Americanisms.* Chicago: University of Chicago Press, 1956.
McCulloch, Walter F. *Woods Words: A Comprehensive Dictionary of Logging Terms.* Corvallis: Oregon Historical Society, 1977.
Michelson, Gunther. *A Thousand Words of Mohawk.* Ottawa: National Museum of Man, 1973.
Partridge, Eric. *Origins: A Short Etymological Dictionary of Modern English.* London: Routledge & Kegan Paul, 1958.
———. *A Dictionary of Slang and Unconventional English.* New York: Macmillan, 1961.
Pokagon, Chief Simon. *Ogimawkwe Mitigwaki (Queen of the Woods).* Hartford, Mich.: C. H. Engle, 1899. (Potawatomi.)
Rand, Silas T. *Rand's Micmac Dictionary.* Charlottetown, Prince Edward Island: Patriot Publishing Co., 1902.
Riggs, Stephen R. *A Dakota-English Dictionary.* Minneapolis: Ross & Haines, 1968.
Skeat, Walter W. *The Language of Mexico, and Words of West Indian Origin.* London: The Philological Society, 1890.
———. *Etymological Dictionary of the English Language.* Oxford: Clarendon Press, 1911; 4th ed., London: Oxford University Press, 1961.
Strachey, William. "A Dictionarie of the Indian Language." In *The Historie of Travaile into Virginia Britannia,* 185–96. London: Hakluyt Society, 1845.
Trumbull, James H. *Natick Dictionary.* Bureau of American Ethnology Bulletin no. 25. Washington, D.C.: U.S. Government Printing Office, 1903.
Watkins, E. A. *A Dictionary of the Cree Language.* Toronto: Anglican Book Center, 1981.
*Webster's New World Dictionary of the American Language.* 2d collegiate ed. N.p.: William Collins & World Publishing Co., 1978.
Williams, Roger. *A Key into the Language of America.* Reprint. Ann Arbor: Gryphon Press, 1971.
Williamson, John P. *An English-Dakota Dictionary.* Minneapolis: Ross & Haines, 1970.

## Books

Adams, James N. *Illinois Place Names.* Springfield: Illinois State Historical Society, 1969. Reprint from *Illinois Libraries* 50, nos. 4, 5, 6 (1968).
Akrigg, G. P. V., and Helen Akrigg. *1001 British Columbia Place Names.* Vancouver: Discovery Press, 1973.
Armour, David A., ed. *Attack at Michilimackinac, 1763.* Mackinac Island: Mackinac Island State Park Commission, 1971. (Travels of Alexander Henry.)
Atwater, Caleb. *The Indians of the Northwest, Their Manners, Customs, &c. or Remarks Made on a Tour to Prairie du Chien. . . .* Columbus: 1850.
Baker, Ronald, and Marvin Carmony. *Indiana Place Names.* Bloomington: Indiana University Press, 1975.

Barnett, Louise. *Ignoble Savage.* Westport, Conn.: Greenwood Press, 1975.

Baxter, Albert. *History of the City of Grand Rapids, Michigan.* New York & Grand Rapids: Munsell & Co., 1891.

Bean, Lowell, and Harry Lawton. *The Cahuila Indians of Southern California.* Banning, Calif.: Malki Museum Press, 1979.

Beauchamp, William M. *Aboriginal Place Names of New York.* Albany: New York State Museum, 1907. Reprint. Detroit: Grand River Books, 1971.

Bissell, Benjamin. *The American Indian in English Literature of the Eighteenth Century.* New Haven: Yale University Press, 1925. Reprint. Hamden, Conn.: Archon Books, 1968.

Blackbird, Andrew J. *History of the Ottawa and Chippewa Indians of Michigan.* 1887. Reprint. Petoskey: Little Traverse Regional Historical Society, 1977.

Blair, Emma H., ed. *The Indian Tribes of the Upper Mississippi and Region of the Great Lakes.* 2 vols. Cleveland: Arthur H. Clark Co., 1911. Reprint. Kraus Reprint Co., 1969.

Bolton, Reginald P. *Indian Life of Long ago in the City of New York.* New York: Crown Publishers, 1972.

Brinton, Daniel G. *The Myths of the New World.* 2d ed. New York: Henry Holt, 1876.

Buechner, Cecilia. *The Pokagons.* Indianapolis: Indiana Historical Society, 1933.

Burrage, Henry S., ed. *Early English and French Voyages 1534–1608.* New York: Charles Scribner's, 1901. Reprint. New York: Dover Publications, n.d.

Campbell, Thomas. *The Poetical Works of Thomas Campbell.* Philadelphia: Lea & Blanchard, 1845.

Carlson, Helen S. *Nevada Place Names.* Reno: University of Nevada Press, 1974.

Carter, James L., and Ernest H. Rankin, eds. *North to Lake Superior. The Journal of Charles W. Penny, 1840.* Marquette: John M. Longyear Research Library, 1970.

Carver, Jonathan. *Travels Through the Interior Parts of North America.* Reprint. Minneapolis: Ross & Haines, 1956.

Cass, Lewis. *Remarks on the Condition, Character, and Languages of the North American Indians.* Boston: Cummings, Hilliard & Co., 1826. (Extract from *North American Review* 50 [January, 1826]).

Catton, Bruce. *Michigan, a Bicentennial History.* New York: W. W. Norton, 1976.

Charlevoix, Pierre Francois Xavier de. *Journal of a Voyage to North America.* 2 vols. N.p.: Readex Microprint, 1966.

———. *History and General Description of New France,* ed. John G. Shea. 6 vols. New York: John G. Shea, 1872.

Chastellux, Marquis de. *Travels in North America.* 2 vols. Chapel Hill: University of North Carolina Press, 1963.

Cleland, Charles F. *A Brief History of Michigan's Indians.* Lansing: Michigan History Division, Michigan Department of State, 1975.

Clifton, James A. *The Prairie People.* Lawrence: Regents Press of Kansas, 1971.

———. *The Pokagons, 1683–1983.* Lanham, Md.: University Press of America, 1984.

Colden, Cadwallader. *The History of the Five Indian Nations.* 2 vols. New York: New Amsterdam Book Co., 1902.

Coolidge, Orville W. A *Twentieth Century History of Berrien County, Michigan.* Chicago: Lewis Publishing Co., 1906.

Coues, Elliott, ed. *The History of the Lewis and Clark Expedition.* 3 vols. Reprint. New York: Dover Publications, 1965.

Cooper, James Fenimore. *The Last of the Mohicans.* New York: Grosset & Dunlap, n.d.

Crevecoeur, Michel-Guillaume St. Jean de. *Journey into Northern Pennsylvania and the State of New York.* Ann Arbor: University of Michigan Press, 1964.

Culin, Stewart. *Games of the North American Indians.* Reprint. New York: Dover Publications, 1975.

Danziger, Edmund J. *The Chippewas of Lake Superior.* Norman: University of Oklahoma Press, 1978.

Dickinson, Lora T. *The Story of Winnetka.* Winnetka, Ill.: Winnetka Historical Society, 1956.

Dillenback, J. D., and Leavitt. *History and Directory of Kent County, Michigan.* Grand Rapids: J. D. Dillenback, 1870.

Dockstader, Frederick J. *Great North American Indians.* New York: Van Nostrand Reinhold, 1977.

Donehoo, George P. *Indian Villages and Place Names in Pennsylvania.* Reprint. Baltimore: Gateway Press, 1977.

Douglas-Lithgow, R. A. *A Dictionary of American-Indian Place and Proper Names in New England.* Salem, Mass.: Salem Press, 1909.

Dunn, Jacob P. *True Indian Stories, with Glossary of Indiana Indian Names.* Indianapolis: Sentinel Publishing Co., 1909. Reprint. North Manchester, Ind.: Lawrence A. Schultz, 1964.

Eastman, Mary. *Dahcotah, or Life and Legends of the Sioux.* New York: John Wiley, 1849.

Eckstorm, Fannie H. *Indian Place-Names of the Penobscot Valley and the Maine Coast.* Orono, Maine: University Press, 1941.

Edmunds, R. David. *The Potawatomis, Keepers of the Fire.* Norman: University of Oklahoma Press, 1978.

Everett, Franklin. *Memorials of the Grand River Valley.* Chicago: The Legal News Co., 1878.

Federal Writers Program (FWP). *Michigan: A Guide to the Wolverine State.* New York: Oxford University Press, 1956.

______. *North Dakota: A Guide to the Northern Prairie State.* Fargo: Knight Printing Co., 1938.

Fitzpatrick, Lilian L. *Nebraska Place Names.* Lincoln: University of Nebraska Press, 1960.

Funk, Wilfred. *Word Origins and their Romantic Stories.* New York: Wilfred Funk, 1950.

Galloway, William A. *Old Chillicothe, Shawnee and Pioneer History.* Xenia, Ohio: Buckeye Press, 1934.

Gannett, Henry. *American Names.* Reprint. Washington: Public Affairs Press, 1947.

Gard, Robert E., and L. G. Sorden. *The Romance of Wisconsin Place Names.* New York: October House, 1968.

Geographic Board of Canada. *Place-Names of Manitoba.* Ottawa: Department of Interior, 1933.

Gheerbrant, Alain, ed. *The Incas: Royal Commentaries of the Inca Garcilaso de la Vega.* New York: Orion Press, 1961.

Gilmore, Melvin R. *Uses of Plants by the Indians of the Missouri River Region.* Reprint. Lincoln: University of Nebraska Press, 1977.

Greenman, Emerson. *The Indians of Michigan.* Lansing: Michigan Historical Commission, 1961.

Gringhuis, Dirk. *Lore of the Great Turtle.* Mackinac Island, Mich.: Mackinac Island State Park Commission, 1970.

Gudde, Erwin. *1000 California Place Names.* Berkeley: University of California Press, 1959.

Haines, Elijah. *The American Indian.* Reprint. Evansville, Ind.: Unigraphic, 1977.

Hale, Horatio, ed. *The Iroquois Book of Rites.* Toronto: University of Toronto Press, 1963.

Hamilton, W. B. *Macmillan Book of Canadian Place Names.* Toronto: Macmillan of Canada, 1978.

Harder, Kelsie. *Illustrated Dictionary of Place Names.* New York: Van Nostrand Reinhold, 1976.

Harrington, John P. *Our State Names.* Washington, D.C.: Smithsonian Institution Press, 1955.

Heath, Mary Frances. *Early Memoirs of Saugatuck, Michigan, 1831–1930.* Grand Rapids: William B. Eerdmans Publishing Co., 1930?

Heckewelder, John G. *History, Manners and Customs of the Indian Nations. . . .* Philadelphia: Historical Society of Pennsylvania, 1876. Reprint. New York: Arno Press, 1981.

———. *Narrative of the Mission of the United Brethren among the Delaware and Mohegan Indians.* Cleveland: Burrows Brothers, 1907. (Contains place-name appendix.)

Hennepin, Louis. *A Description of Louisiana.* New York: John G. Shea, 1880. Reprint. Readex Microprint, n.d.

———. *A New Discovery of a Vast Country in America.* 2 vols. Chicago: A. C. McClurg Co., 1903.

Henry, Alexander. *Travels and Adventures.* Ann Arbor: University Microfilms, 1966.

Hodge, Frederick W., ed. *Handbook of American Indians North of Mexico.* Bureau of American Ethnology Bulletin no. 30. 2 vols. Washington, D.C.: U.S. Government Printing Office, 1907–10.

Houghton, Louise S. *Our Debt to the Red Men: The French-Indians in the Development of the United States.* Boston: Stratford Co., 1918.

Huden, John C. *Indian Place Names of New England.* New York: Museum of the American Indian, 1962.

Hulbert, Archer B. *Portage Paths: The Keys to the Continent.* Vol. 7, *Historic Highways of America.* Cleveland: Arthur H. Clark, 1903.

Humboldt, Alexander von. *Personal Narrative of Travels to the Equinoctial Regions of America. . . .* 3 vols. London: George Bell & Sons, 1877.

Jeness, Diamond. *The Indians of Canada.* 7th ed. Toronto: University of Toronto Press, 1977.

Kane, Joseph N. *The American Counties.* 3d ed. Metuchen, N.J.: Scarecrow Press, 1972.

Kappler, Charles J., ed. *Indian Treaties, 1778–1883.* Reprint. New York: Interland Publishing Co., 1972.

Keating, William. *Narrative of an Expedition to the Source of St. Peter's River. . . .* 2 vols. in one. Reprint. Minneapolis: Ross & Haines, 1959.

Keiser, Albert. *The Indian in American Literature.* New York: Oxford University Press, 1933.

Kellogg, Louise P., ed. *Early Narratives of the Northwest, 1634–1699.* Reprint. New York: Barnes & Noble, 1967.

______. *The French Regime in Wisconsin and the Northwest.* Madison: State Historical Society of Wisconsin, 1925.

Kelton, Dwight H. *Indian Names of Places Near the Great Lakes.* Detroit: Detroit Free Press Printing Co., 1888.

______. *Indian Names and History of the Sault Ste. Marie Canal.* Detroit: Detroit Free Press Printing Co., 1889.

Kenny, Hamill. *West Virginia Place Names, Their Origin and Meaning.* Piedmont, W. Va.: Place Name Press, 1945.

Kenton, Edna, ed. *Black Gowns and Redskins.* London: Longmans Green Co., 1956.

Kinietz, W. Vernon. *The Indians of the Western Great Lakes 1615–1760.* Ann Arbor: University of Michigan Press, 1965.

Kinzie, Mrs. John H. (Juliette). *Wau-Bun, or the 'Early Day' in the Northwest.* Chicago: Rand McNally, 1901.

Kohl, J. G. *Kitchi-Gami.* (London: 1860).

Kroeber, A. L. *Handbook of the Indians of California.* Reprint. New York: Dover Publications, 1976.

Lahontan, Louis Armand, Baron de. *New Voyages to North America.* Edited by R. G. Thwaites. 2 vols. Chicago: A. C. McClurg, 1905.

Landes, Ruth. *Ojibwa Religion and the Midewiwin.* Madison: University of Wisconsin Press, 1968.

Lanman, James H. *History of Michigan.* New York: E. French, 1839.

Lavender, David. *Winner Take All: The Trans-Canada Canoe Trail.* New York: McGraw Hill, 1977.

Lawson, John. *History of North Carolina.* Reprint. Richmond: Garrett & Massie, 1937.

Leeson, M. A. *History of Kent County, Michigan.* Chicago: C. C. Chapman Co., 1881.

Lewis, Ferris E. *Michigan Yesterday and Today.* 6th ed. Hillsdale, Mich.: Hillsdale Educational Publishers, 1967.

Lillie, Leo C. *Historic Grand Haven and Ottawa County.* Grand Rapids: A. P. Johnson, 1931.

McKenney, Thomas L. *Sketches of a Tour to the Lakes.* Barre, Mass.: Imprint Society, 1972.

McReynolds, Edwin C. *The Seminoles.* Norman: University of Oklahoma Press, 1957.

Mathews, Alfred. *History of Cass County, Michigan.* Chicago: Waterman, Watkins & Co., 1882.

Meany, Edmond S. *Origin of Washington Geographic Names.* Seattle: University of Washington Press, 1923. Reprint. Detroit: Gale Research Co., 1968.

Morgan, Lewis H. *Ancient Society.* Chicago: Charles H. Kerr Co., [1910].

———. *League of the Ho-de-no-sau-nee or Iroquois.* Reprint. New Haven: Human Relations Area Files, 1954.

Morse, Jedidiah. *Report to the Secretary of War on Indian Affairs.* New Haven: S. Converse, 1822.

Myers, Albert C., ed. *Narratives of Early Pennsylvania, West Jersey and Delaware, 1630–1707.* Reprint. New York: Barnes & Noble, 1959.

Nicollet, Joseph N. *Report Intended to Illustrate a Map of the Hydrographic Basin of the Upper Mississippi River,* 26th Cong., 2d sess., S. Doc. 237. Washington: Blair & Rives, 1843.

———. *The Journals of Joseph N. Nicollet.* Reprint. St. Paul: Minnesota Historical Society, 1970.

Oliver, David D. *Centennial History of Alpena County, Michigan, From 1837 to 1876,* Alpena: Argus Printing House, 1903.

Orth, Donald J. *Dictionary of Alaska Place Names.* U.S. Geological Survey Professional Paper no. 567. Washington, D.C.: U.S. Government Printing Office, 1967.

Osborn, Chase S. and Stellanova. *Schoolcraft, Longfellow, Hiawatha.* Lancaster, Pa.: Jacques Cattell Press, 1942.

Oviedo, Gonzalo Fernandez de. *Natural History of the West Indies.* Reprint. Chapel Hill: University of North Carolina Press, 1959.

Page, H. R. and Co. *History of Manistee, Mason and Oceana Counties.* Chicago: 1882.

Parkman, Francis. *The Jesuits in North America in the Seventeenth Century.* Boston: Little Brown, 1896.

———. *The Conspiracy of Pontiac.* 2 vols. Boston: Little Brown, 1922.

Peckham, Howard H. *Pontiac and the Indian Uprising.* Princeton: Princeton University Press, 1947. Reprint. Chicago: University of Chicago Press, 1961.

Peters, Bernard C., ed. *Lake Superior Journal: Bela Hubbard's Account of the 1840 Houghton Expedition.* Marquette: Northern Michigan University Press, 1983.

Phillips, James W. *Alaska-Yukon Place Names.* Seattle: University of Washington Press, 1973.

Powers, Stephen. *Tribes of California.* Reprint. Berkeley: University of California Press, 1976.

Prescott, William H. *History of the Conquest of Mexico and History of the Conquest of Peru.* New York: Modern Library, n.d.

Quaife, Milo M. *Lake Michigan.* Indianapolis: Bobbs Merrill, 1944.

———. *The Western Country in the Seventeenth Century.* New York: Citadel Press, 1962.

Radin, Paul. *The Winnebago Tribe.* Reprint. Lincoln: University of Nebraska Press, 1970.

Read, William A. *Louisiana Place Names of Indian Origin.* Bulletin 19. Baton Rouge: Louisiana State University, 1927.

———. *Florida Place Names of Indian Origin.* Baton Rouge: Louisiana State University Press, 1934.

———. *Indian Place-Names in Alabama.* Baton Rouge: Louisiana State University Press, 1937.

Reber, L. Benjamin. *Early History of St. Joseph.* St. Joseph: Chamber of Commerce, n.d.

Richardson, John. *Wacousta, or The Prophesy.* 2 vols. New York: Dewitt & Davenport, 1832.

Rogers, Howard S. *History of Cass County, Michigan, from 1825 to 1875.* Cassopolis: W. H. Mansfield, 1875.

Rogers, Robert. *Journal of Maj. Robert Rogers.* Ann Arbor: University Microfilms, 1966.

———. *A Concise Account of North America.* Reprint. New York: Johnson Reprint Corporation, 1967.

Romig, Walter. *Michigan Place Names.* Grosse Pointe: Walter Romig, n.d.

Rosalita, Sister M., I.H.M. *Detroit: The Story of Some Street Names.* Detroit: Wayne State University Press, 1951.

Rourke, Constance. *The Roots of American Culture.* New York: Harcourt Brace, 1942.

Ruttenber, E. M. *Footprints of the Red Men.* Proceedings New York State Historical Association vol. 6. Albany: 1906.

Rydjord, John. *Kansas Place-Names.* Norman: University of Oklahoma Press, 1972.

Schoolcraft, Henry R. *Narrative Journal . . . in the Year 1820.* Albany: E. & F. Hosford, 1821.

———. *Notes on the Iroquois.* Albany: Erastus H. Pease, 1847.

———. *Information Respecting the History, Condition, and Prospects of the Indian Tribes of the United States.* 6 vols. Philadelphia: Lippincott, Grambo, 1851–56.

———. *Personal Memoirs of a Residence of Thirty Years with the Indian Tribes. . . .* Reprint. New York: Arno Press, 1975.

———. *Legends of the American Indians.* Reprint. New York: Crescent Books, 1980. Reprint of selections from *Algic Researches.*

Schwarz, Herbert T. *Windigo and Other Tales of the Ojibways.* Toronto: McClelland Stewart Ltd., 1972.

Scott, Irving D. *Inland Lakes of Michigan.* Lansing: Wynkoop, Hallenbeck, Crawford Co., 1921.

Smith, C. Henry. *Metamora.* Bluffton, Ohio: College Book Store, 1947.

Sneve, Virginia Driving Hawk. *South Dakota Geographic Names.* Sioux Falls: Brevet Press, 1973.

Sprague, Elvin L., and Mrs. George N. Smith. *Sprague's History of Grand Traverse and Leelanau Counties.* N.p.: B. F. Bowen, 1903.

Stewart, Catherine. *New Homes in the West.* Ann Arbor: University Microfilms, 1966.

Stewart, George R. *American Place Names.* New York: Oxford University Press, 1970.

Swanton, John R. *The Indian Tribes of North America.* Bureau of American Ethnology Bulletin no. 145. Washington, D.C.: U.S. Government Printing Office, 1952.

Tanner, John. *A Narrative of the Captivity and Adventures of John Tanner.* Edited by Edwin C. James. Reprint. Minneapolis: Ross & Haines, 1956.

Thoreau, Henry D. *The Maine Woods.* Cambridge: Riverside Press, 1906.

Thwaites, Reuben G., ed. *Jesuit Relations and Allied Documents.* 73 vols. Cleveland: Burrows Brothers, 1896–1901.

———, ed. *Early Western Travels.* 32 vols. Cleveland: Arthur H. Clark, 1904–7.

Tooker, Elisabeth. *An Ethnography of the Huron Indians, 1615–1649.* Bureau of American Ethnology Bulletin no. 190. Washington, D.C.: U.S. Government Printing Office, 1964.

Tooker, William W. *The Indian Place Names on Long Island.* Reprint. Port Washington, N.Y.: Ira J. Friedman, 1962.

Trumbull, James H. *Indian Names in Connecticut.* Reprint. Hamden, Conn.: Archon Books, 1974.

Tucker, Glenn. *Tecumseh: Vision of Glory.* Indianapolis: Bobbs Merrill, 1956.

Tyler, Lyon G., ed. *Narratives of Early Virginia 1606–1625.* Reprint. New York: Barnes & Noble, 1959.

Underhill, Ruth M. *Red Man's Religion.* Chicago: University of Chicago Press, 1965.

Upham, Warren. *Minnesota Geographic Names.* Reprint. St. Paul: Minnesota Historical Society, 1969.

Vogel, Virgil J. *Indian Place Names in Illinois.* Springfield: Illinois State Historical Society, 1963.

———. *American Indian Medicine.* Norman: University of Oklahoma Press, 1970.

———. *This Country Was Ours.* New York: Harper & Row, 1973.

———. *Iowa Place Names of Indian Origin.* Iowa City: University of Iowa Press, 1983.

Wallace, Ernest, and E. Adamson Hoebel. *The Comanches, Lords of the South Plains.* Norman: University of Oklahoma Press, 1952.

Warren, William W. *History of the Ojibway Nation.* Reprint. Minneapolis: Ross & Haines, 1970.

White, Maurice. *Indian Names and Meanings.* Washington, D.C.: Educational Research Bureau, 1946.

Whiting, Henry. *Ontwa, the Son of the Forest.* New York: 1823.

———. *Sannilac, a Poem by Henry Whiting, with Notes by Lewis Cass and Henry R. Schoolcraft.* Boston: Carter, Hendee & Babcock, 1831.

Williams, Mentor. *Schoolcraft's Indian Legends.* East Lansing: Michigan State University Press, 1956. From *Algic Researches.*

Williams, Ralph D. *Honorable Peter White.* N.p.: Penton Publishing Co., n.d.

Winger, Otho. *The Potawatomi Indians.* Elgin, Ill.: Elgin Press, 1939.

## Articles

Ackerknecht, Erwin H. "White Indians. . . ." *Bulletin of the History of Medicine* 15 (January, 1944): 15–36.
Andrus, Percy H. "Markers and Memorials in Michigan." *Michigan History* 15 (Spring, 1931): 167–374.
"Appointment of Augustin Hamelin Jr. as head chief by Ottawas." *Michigan Pioneer and Historical Society Collections* 2 (1887): 621–22.
Armitage, B. Phillis. "A Study of Michigan's Place Names." *Michigan History* 27 (October–December, 1943): 626–37.
Boelio, Bob. "The Islands of Michigan." *Chronicle* 19 (Fall, 1983): 34–39.
______. "The Rivers of Michigan." *Chronicle* 20 (Spring, 1984): 18–21.
Brooks, Harlow. "The Medicine of the American Indians." *Journal of Laboratory and Clinical Medicine* 19 (October, 1933): 1–23.
Brotherton, R. A. "Meaning of Escanaba." *Inland Seas* 4 (Fall, 1948): 210–11.
Browne, C. A. "The Chemical Industries of the American Aborigines." *Isis* 23 (1935): 406–24.
Brunson, Catherine C. "A Sketch of Pioneer Life among the Indians." *Michigan Pioneer and Historical Society Collections* 28 (1897–98): 161–63.
Butler, Albert F. "Rediscovering Michigan's Prairies." *Michigan History* 32 (March, 1948): 15–36.
Campbell, James V. "The Small Perils of History." *Michigan Pioneer and Historical Society Collections* 30 (1905): 396–404.
Carpenter, C. E. "Squaw Island—How it Received its Name." *Michigan Pioneer and Historical Society Collections* 13 (1888): 486–88.
Chamberlain, Alexander F. "Signification of Certain Algonquian Animal Names." *American Anthropologist,* n.s., 3 (1901): 669–83.
______. "Algonkian Words in American English. . . ." *Journal of American Folk-Lore* 15 (October–December, 1902): 240–67.
Childs, Mrs. W. A. "Reminiscences of Old Keweenaw." *Michigan Pioneer and Historical Society Collections* 30 (1905): 150–55.
Christian, E. P. "Historical Associations Connected with Wyandotte and Vicinity." *Michigan Pioneer and Historical Society Collections* 13 (1888): 308–24.
Copley, A. B. "The Pottawattomies." *Michigan Pioneer and Historical Society Collections* 14 (1889): 256–67.
Denby, Charles. "Meaning of the Name Huron as Applied to the Huron Indians." *Michigan History* 13 (July, 1929): 436–42.
Dewey, F. A. "Me-te-au, a King, and Where He Reigned." *Michigan Pioneer and Historical Society Collections* 13 (1888): 567–71.
Dodge, Mrs. Frank P. "Landmarks of Lenawee County." *Michigan Pioneer and Historical Society Collections* 38 (1912): 478–90.
Dunn, Jacob P. "Indiana Geographical Nomenclature." *Indiana Quarterly Magazine of History* 8 (September, 1912): 109–14
______. "Names of the Ohio River." *Indiana Quarterly Magazine of History* 8 (1912): 166–70.

Dustin, Fred. "Some Indian Place-names around Saginaw." *Michigan History* 12 (October, 1928): 729–39.

______. "Isle Royale Place Names." *Michigan History* 30 (October–December, 1946): 681–722.

Eells, Myron. "Aboriginal Geographic Names in the State of Washington." *American Anthropologist* 5 (1892): 27–35.

Emmert, D. G. "The Indians of Shiawassee County." *Michigan History* 47 (September, 1963): 243–72.

Errett, Russell. "Indian Geographical Names, Part II." *Magazine of Western History* 2 (July, 1885): 238–46.

Farrand, Mrs. B. C. "Early Days in Desmond and Vicinity." *Michigan Pioneer and Historical Society Collections* 13 (1888): 334–40.

Felch, Alpheus. "The Indians of Michigan and the Cession of Their Lands to the U.S. by Treaties." *Michigan Pioneer and Historical Society Collections* 26 (1894–95): 274–97.

Ferry, William N. "Ottawa's Old Settlers." *Michigan Pioneer and Historical Society Collections* 30 (1905): 572–82.

Florida Historical Society. "The Complete Story of Osceola." *Florida Historical Quarterly* 33 (January–April, 1955): entire issue.

Forbes, Abbé Joseph William. "St. Francis of Caughnawaga." *Kateri,* no. 122 (Winter, 1979): 27–30.

Foster, Ava. "Indian Names." *Totem Pole* 29 (August, 1952): 1–6.

Foster, Theodore G. "Indian Place Names in Michigan." *Totem Pole* 28 (January, 1952): 3–4.

Fox, George R. "Place Names of Berrien County." *Michigan Historical Magazine* 8 (January, 1924): 6–35.

______. "Place Names of Cass County." *Michigan History Magazine* 27 (Summer, 1943): 463–91.

Gagnieur, William F. "Indian Place Names in the Upper Peninsula, and Their Interpretation." *Michigan History* 2 (July, 1918): 525–55.

______. "Indian Place Names in the Upper Peninsula of Michigan and Elsewhere." *Michigan History* 3 (July, 1919): 412–19.

______. "Indian Place-names." *Michigan History* 9 (January, 1925): 109–11.

______. "Ketekitiganing (Lac Vieux Desert)." *Michigan History* 12 (October, 1928): 776–77.

______. "Tahquamenon." *Michigan History* 14 (July, 1930): 557.

Ganong, William F. "A Monograph on the Place-nomenclature of the Province of New Brunswick." *Royal Society of Canada, Proceedings and Transactions* (1896): 175–289.

Gatschet, A. S. "The Fish in Local Onomatology." *American Anthropologist* 5 (October, 1892): 361–62.

______. "Tecumseh's Name." *American Anthropologist* 8 (1895): 91–92.

Gerard, William R. In "Anthropological Miscellany." *American Anthropologist,* n.s., 13 (April–June, 1911): 337–38.

______. "Plant Names of Indian Origin." *Garden and Forest* 9 (July 15, 1896): 282–83.

______. "Virginia's Indian Contributions to English." *American Anthropologist*, n.s., 9 (1907): 87–112.

______. "Kalamazoo." *American Anthropologist*, n.s., 13 (April–June, 1911): 337–38.

Goss, Dwight. "The Indians of the Grand River Valley." *Michigan Pioneer and Historical Society Collections* 30 (1905): 172–90.

Gray, Martha. "Reminiscences of the Grand Traverse Region." *Michigan Pioneer and Historical Society Collections* 38 (1912): 285–88.

Greenman, Emerson. "Indian Chiefs of Michigan." *Michigan History* 23 (Summer, 1939): 220–49.

Haines, Blanche M. "French and Indian Footprints at Three Rivers on the St. Joseph." *Michigan Pioneer and Historical Society Collections* 38 (1912): 386–97.

Hamilton, Charlotte. "Chippewa County Place Names." *Michigan History* 27 (October–December, 1943): 638–43.

Hartley, Alan H. "The Expansion of Ojibway and French Place-Names into the Lake Superior Region in the Seventeenth Century." *Names* 28 (March, 1980): 43–68.

Hoffman, Walter J. "The Midewiwin or 'Grand Medicine Society' of the Ojibwa." *Seventh Annual Report, Bureau of American Ethnology*. Washington, D.C.: U.S. Government Printing Office, 1891.

______. "The Menomini Indians." *Fourteenth Annual Report, Bureau of American Ethnology*, pt. 1. Washington, D.C.: U.S. Government Printing Office, 1896.

Holmer, Nils M. "Indian Place Names in South America and the Antilles, Part I." *Names* 8 (September, 1960): 133–49.

______. "Indian Place Names in South America and the Antilles, Part II." *Names* 8 (December, 1960): 197–219.

Holmes, William H. "The Tomahawk." *American Anthropologist*, n.s., 10 (1908): 264–76.

Hooker, John J. "Daniel Marsac," and "John S. Hooker of Lowell." *Michigan Pioneer and Historical Society Collections* 38 (1912): 60–64.

Hoyt, James M. "History of the Town of Commerce." *Michigan Pioneer and Historical Society Collections* 14 (1889): 421–30.

"Indian Names." *Michigan Pioneer and Historical Society Collections* 7 (1884): 136.

Jenks, Albert E. "The Wild Rice Gatherers of the Upper Lakes." *Nineteenth Annual Report, Bureau of American Ethnology*, pt. 2. Washington, D.C.: U.S. Government Printing Office, 1901.

Jenks, William L. "History and Meaning of the County Names of Michigan." *Michigan Pioneer and Historical Society Collections* 38 (1912): 439–78. Summarized in *Michigan History* 10 (October, 1926): 646–55.

Jenks, William L. "Some Early Maps of Michigan." *Michigan Pioneer and Historical Society Collections* 38 (1912): 627–37.

Johnson, William W. "Indian Names in the County of Mackinac." *Michigan Pioneer and Historical Society Collections* 12 (1887): 375–81.

Kelton, Dwight H. "Mackinac County." *Michigan Pioneer and Historical Society Collections* 6 (1883): 343–56.

Kingsbury, Stewart A. "Sets and Name Duplication in the Upper Peninsula of Michigan." *Names* 29 (December, 1981): 307–12.

Kinzie, Juliette A. "Chicago Indian Chiefs." *Bulletin of Chicago Historical Society* 1 (August, 1935): 105–16.

Kroeber, A. L. "California Place Names of Indian Origin." *University of California Publications in American Archaeology and Ethnology* 12 (June 15, 1916): 31–69.

Kuhm, Herbert W. "Indian Place-Names in Wisconsin." *Wisconsin Archeologist*, n.s., 33 (March and June, 1952): entire issue.

Lemmer, Victor F. "History of the County of Gogebic, 1965." *Courthouse Review* (December, 1965–January, 1966). Reprint.

Mahr, August C. "Indian River and Place Names in Ohio." *Ohio Historical Quarterly* 66 (1957): 137–58.

McCormick, William R. "Indian Stoicism and Courage." In H. R. Page Co., *History of Bay County, Michigan* (n.p.: 1883).

———. "A Pioneer Incident." *Michigan Pioneer and Historical Society Collections* 4 (1883): 376–79.

———. "Indian Names in the Saginaw Valley." *Michigan Pioneer and Historical Society Collections* 7 (1884): 277.

McMullen, E. Wallace. "Prairie Generics in Michigan." *Names* 7 (September, 1959): 188–90.

Marckwardt, Albert H. "Naming Michigan's Counties." *Names* 23 (September, 1975): 180–89.

Martin, Maria Ewing. "Origin of Ohio Place Names." *Ohio Archaeological and Historical Society Publications* 14 (1905): 272–90.

May, George S. "The Meaning and Pronunciation of Michilimackinac." *Michigan History* 42 (December, 1958): 385–90.

*Michigan History Magazine* 42 (special Mackinac issue) (December, 1958).

Miller, Albert. "Incidents of Early Saginaw." *Michigan Pioneer and Historical Society Collections* 13 (1889): 351–79.

———. "The Rivers of the Saginaw Valley Sixty Years Ago." *Michigan Pioneer and Historical Society Collections* 14 (1889): 495–510.

Moore, V. L. "Baw Beese Lake." *Michigan History* 16 (Spring, 1932): 334–47.

Niles, Mrs. M. J. "Sketch of Old Times in Clinton County." *Michigan Pioneer and Historical Society Collections* 14 (1889): 620–26.

Norris, L. D. "History of Washtenaw County." *Michigan Pioneer and Historical Society Collections* 1 (1874–76): 327–33.

Osband, Melvin D. "The Story of Tonguish." *Michigan Pioneer and Historical Society Collections* 8 (1885): 161–64.

Peters, Bernard C. "The Origin of Some Stream Names along Michigan's Lake Superior Shoreline." *Inland Seas* 37 (Spring, 1981): 6–12.

———. "The Origin and Meaning of Place Names along Pictured Rocks National Lakeshore." *Michigan Academician* 14 (Summer, 1981): 41–55.

———. "The Origin and Meaning of Chippewa Place Names along the Lake Superior Shoreline." *Names* 22 (September, 1984): 234–51.

Petoskey, Ella. "Chief Petoskey." *Michigan History* 13 (July, 1929): 443–48.

Poppleton, O. "How Battle Creek Received its Name." *Michigan Pioneer and Historical Society Collections* 6 (1883): 248–51.

Ransom, J. Ellis. "Derivation of the Word Alaska." *American Anthropologist*, n.s., 42 (1940): 550–51.
Rayburn, J. A. "Geographical Names of Amerindian Origin in Canada, Part II." *Names* 17 (June, 1969): 49–58.
Read, Allen W. "The Rationale of Podunk." *American Speech* 14 (April, 1939): 99–108.
"Reports of Counties, Towns and Districts." *Michigan Pioneer and Historical Society Collections* 1 (1874–75): 94–520.
St. John, Mrs. S. E. "Daily Life, Manners and Customs of the Indians of Kalamazoo County," *Michigan Pioneer and Historical Society Collections* 10 (1887): 170–72. (Undated excerpt from *Kalamazoo Telegraph.*)
Scott, A. H. "Indians in Kalamazoo County." *Michigan Pioneer and Historical Society Collections* 10 (1887): 163–66.
Shout, Mary E. "Reminiscences of the First Settlement at Owosso." *Michigan Pioneer and Historical Society Collections* 30 (1905): 344–52.
Smith, C. Henry. Letter on the origin of Metamora. *Michigan History* 28 (April–June, 1944): 319–20.
Smith, Edwin S. "Pioneer Days in Kalamazoo and Van Buren." *Michigan Pioneer and Historical Society Collections* 14 (1889): 272–80.
Smith, Emerson R. "Michilimackinac, Land of the Great Fault." *Michigan History*, 42 (December, 1958): 392–95.
Smith, Harlan I. "The Invasion of the Saginaw Valley." *Michigan Pioneer and Historical Society Collections* 28 (1897–98): 642–45.
Snow, Vernon F. "From Ouragan to Oregon." *Oregon Historical Quarterly* 60 (December, 1959): 439–47.
Spooner, Harry L. "Indians of Oceana." *Michigan History* 15 (Autumn, 1931): 654–65.
Stewart, George R. "The Source of the Name 'Oregon.'" *American Speech* 19 (April, 1944): 115–17.
______. "Ouaricon Revisited." *Names* 15 (September, 1967): 166–68.
Sutton, George. "An Old Time Murder in Northfield." *Michigan Pioneer and Historical Society Collections* 17 (1890): 511–12.
"Topinabee." *Totem Pole* 28 (March 3, 1952): 1–3.
Thwaites, Reuben G., ed. "The French Regime in Wisconsin." *Collections State Historical Society of Wisconsin* 18 (1908).
Trowbridge, C. C. "Meearmeear Traditions." *Occasional Contributions, University of Michigan Museum of Anthropology* 7 (1938): 1–191.
______. "Shawnese Traditions." *Occasional Contributions, University of Michigan Museum of Anthropology* 9 (1939): 1–76.
Trumbull, J. H. "The Composition of Indian Geographical Names, Illustrated from the Algonkin Languages." *Collections Connecticut Historical Society* 2 (1870): 3–50.
Turner, Jesse. "Reminiscences of Kalamazoo." *Michigan Pioneer and Historical Society Collections* 17 (1890): 570–88.
Turrell, Archie M. "Some Place Names of Hillsdale County." *Michigan History* 6, no. 4 (Fall, 1922): 573–82.
Van Buren, A. D. P. "Deacon Isaac Mason's Early Recollections of Michigan." *Michigan Pioneer and Historical Society Collections* 5 (1882): 397–402.

______. "Indian Reminiscences of Calhoun and Kalamazoo Counties." *Michigan Pioneer and Historical Society Collections* 10 (1887): 147–63.

______. "The First Settlers in the Township of Battle Creek." *Michigan Pioneer and Historical Society Collections* 5 (1882): 292–93.

______. "Story of the Baw Beese Indians." *Michigan Pioneer and Historical Society Collections* 28 (1897–98): 530–33.

Vaudreuil, Pierre de. "Commission to the King of Monguagon." *Michigan Pioneer and Historical Society Collections* 8 (1885): 459.

Verwyst, Chrysostom. "Geographical Names in Wisconsin, Minnesota, and Michigan Having a Chippewa Origin." *Collections State Historical Society of Wisconsin* 12 (1892): 390–98.

______. "A Glossary of Chippewa Indian Names of Rivers, Lakes and Villages." *Acta et Dicta* 4 (July, 1916): 255–74.

Vogel, Virgil J. "The Missionary as Acculturation Agent: Peter Dougherty and the Indians of Grand Traverse." *Michigan History* 51 (Fall, 1967): 185–201.

______. "Oregon, a Rejoinder." *Names* 16 (June, 1968): 136–40.

______. "American Indian Foods Used as Medicine." In *American Folk Medicine,* edited by Wayland C. Hand, 125–41. Berkeley: University of California Press, 1976.

Waite, Minnie B. "Indian and Pioneer Life." *Michigan Pioneer and Historical Society Collections* 38 (1912): 318–21.

Walton, Ivan H. "Indian Place Names in Michigan." *Midwest Folklore* 5 (Spring, 1955): 23–34.

______. "Origin of Names on the Great Lakes." *Names* 3 (December, 1955): 239–46.

Webber, William L. "Indian Cession of 1819, Made by the Treaty of Saginaw." *Michigan Pioneer and Historical Society Collections* 26 (1894–95): 517–34.

Weissert, Charles A. "The Indians of Barry County and the Work of Leonard Slater, the Missionary." *Michigan History* 16, no. 3 (Summer, 1932): 321–33.

______. "Indians in 'Bitter Minority' in District until Expulsion." *Michigan Pioneer and Historical Society Collections* 8 (1885): 459.

White, George H. "Yankee Lewis's Famous Hostelry in the Wilderness." *Michigan Pioneer and Historical Society Collections* 26 (1894–95): 302–7.

White, Peter. "The Iron Region of Lake Superior." *Michigan Pioneer and Historical Society Collections* 8 (1885): 145–60.

"White Pigeon's Grave." *Michigan Pioneer and Historical Society Collections* 10 (1887): 172–74. (Extract from *White Pigeon Republican,* May 29, 1839.)

Williams, Ephraim. "Personal Reminiscences." *Michigan Pioneer and Historical Society Collections* 8 (1885): 233–59.

______. "Remembrances of Early Days." *Michigan Pioneer and Historical Society Collections* 10 (1886): 137–47.

______. "What I Know about O-taw-wars and Ne-war-Go." *Michigan Pioneer and Historical Society Collections* 7 (1884): 134–50.

Williams, R. V. "Washtenaw County." *Michigan Pioneer and Historical Society Collections* 4 (1881): 393–94.

Wilson, Etta Smith. "Life and Work of the Late George N. Smith, Pioneer Missionary." *Michigan Pioneer and Historical Society Collections* 30 (1905): 190–212.

Zeisberger, David. "A History of the Indians." *Ohio Archaeological and Historical Quarterly* 19 (January & April, 1910): 12–153.

## Local, State, and Federal Government Documents

Carter, Clarence Edward, ed. *Michigan, 1805–20. Michigan, 1829–37. Illinois, 1809–14.* Vols. 10, 12, 16, *The Territorial Papers of the United States.* Washington, D.C.: U.S. Government Printing Office, 1934–58.

Michigan Commission on Indian Affairs. *Annual Report for 1981.* Lansing: 1982.

———. *Michigan Indian Directory.* Lansing: 1981.

Michigan Townships Association. *Michigan Township Officials: Directory 1981–84.* Lansing: n.d.

Taylor, Theodore W. *The States and their Indian Citizens.* Washington, D.C.: U.S. Department of Interior, 1972.

United States Department of Commerce. *Federal and State Indian Reservations.* Washington, D.C.: 1971.

United States House of Representatives. *Report with Respect to the House Resolution Authorizing . . . an Investigation of the Bureau of Indian Affairs, December 15, 1952.* Washington, D.C.: U.S. Government Printing Office, 1953.

## Atlases, Gazetteers, and Maps

Board of County Road Commissioners. *Official Map of Baraga County.* L'Anse: n.d.

Blois, John T. *Gazetteer of the State of Michigan.* Detroit: Sydney L. Rood & Co., 1838. Reprint. New York: Arno Press, 1975.

Gogebic County Board of Commissioners, et al. *Gogebic County Map.* Bessemer: n.d.

Karpinski, Louis C. *Historical Atlas of the Great Lakes and Michigan.* Lansing: Michigan Historical Commission, 1931.

Michigan State Highway Department. *Official Road Map of Michigan.* Issued annually.

Michigan Tourist Council, in Cooperation with Michigan Department of Conservation. *Michigan Canoe Trails.*

Mid-Michigan Map Service. *Map of Greater Lansing.*

Rand McNally & Co. *Commercial Atlas and Marketing Guide.* Chicago: Rand McNally & Co., 1981.

———. *Road Atlas and Travel Guide.* Chicago: Rand McNally & Co., 1982.

Rockford Map Publishers. *Michigan State Atlas.* Rockford, Ill.: Rockford Map Publishers, 1980.

Tucker, Sara Jones. *Indian Villages of the Illinois Country.* Vol. 2, *Scientific Papers.* Pt. 1, *Atlas.* Springfield: Illinois State Museum, 1942.

United States Army Corps of Engineers. General Chart of the Great Lakes, 1955.

United States Forest Service, Department of Agriculture. Maps of various national forests.

United States Geological Survey, Department of Interior. Topographic Maps of Upper and Lower Michigan, and local quadrants.

## References and Guides

Barnhart, Clarence, ed. *New Century Cyclopedia of Names.* 3 vols. New York: Appleton Century Crofts, 1954.

National Native American Cooperative. *Native American Directory.* San Carlos, Ariz.: 1982.

National Railway Publication Co. *The Official Guide to the Railways and Steam Navigation Lines of the United States.* New York: June, 1958.

———. *The Official Railway Guide, North American Freight Service Edition.* New York: May–June, 1979.

Nichols, Frances S. *Index to Schoolcraft's "Indian Tribes of the United States."* Bureau of American Ethnology Bulletin no. 152. Washington, D.C.: U.S. Government Printing Office, 1954.

Sealock, Richard B., Margaret M. Sealock, and Margaret S. Powell. *Bibliography of Place-Name Literature, United States and Canada.* 3d ed. Chicago: American Library Association, 1982.

United States Postal Service. *Directory of Post Offices.* Washington, D.C.: 1960, 1974, 1982.

## Newspaper Items

*Chicago Daily News,* Beeline, January 22, 1978 (Podunk).

———. February 7, 1978 (Michigan Indians today).

*Kalamazoo Gazette,* December 29, 1946, p. 10. "Indians in 'Bitter Minority' in District until Expulsion."

*Marquette Mining Journal,* undated clips, 1902 (Kawbawgam death).

*New York Times,* October 18, 1970, April 3, 1978 (Michigan Indians today).

# Index